D0384807

Faces of Science

MARIANA COOK

Faces of Science

INTRODUCTION BY GERARD PIEL

W. W. NORTON & COMPANY
NEW YORK LONDON

Printed in Italy
First Edition

The text of this book is composed in Bodoni book and Bauer Bodoni
Book design and composition by Katy Homans
Manufacturing by Mondadori Printing, Verona

Library of Congress Cataloging-in-Publication Data

Cook, Mariana Ruth.
Faces of science : portraits / Mariana Cook ; introduction by Gerard Piel.— 1st ed.
p. cm.
Includes index.
ISBN 0-393-06118-3 (hardcover)
1. Scientists—Portraits. 2. Portrait photography. 3. Cook, Mariana Ruth. 4. Scientists—Biography. I. Title.
TR681.S37C66 2005
509.2'2—dc22

2005006372

ISBN: 0-393-06118-3

W.W. Norton & Company, 500 Fifth Avenue, New York, NY 10110
www.wwnorton.com
W.W. Norton & Company Ltd., Castle House, 75/76 Wells Street, London, WIT 3QT

1 2 3 4 5 6 7 8 9 0

Contents

To those who ask why

Preface

MARIANA COOK

Over the years, I have been commissioned to photograph a number of scientists. In August of 2001, I made portraits of twenty-four genome scientists at a symposium in Santa Cruz, California. I had individual sessions with each of them over a three-day period and was struck by their extraordinary intelligence and their directness. They spoke of their work with excitement and were not offended by how little I knew about science. They were able to explain their work and their findings clearly and in lay terms. I was also impressed by the realization that twenty-two of the twenty-four scientists had stable personal lives, never divorced after decades of marriage. This struck me as odd because many of them seemed to be quite eccentric. At the same time, they had their feet on the ground. It is an unusual combination.

I wished to photograph more scientists, but did not know how to choose them. In February of 2003, I sat next to Gerard Piel, the former publisher of *Scientific American* magazine, at a dinner. He spoke of many scientists and told me about their work with great enthusiasm. I was inspired and knew he was just the right person to advise me on which scientists to photograph for the book I hoped to make. He said I could call him the next day. That night, I dreamt we met for lunch at a restaurant that had square tables with white paper tablecloths. In the dream, I covered the tablecloth with names of scientists, and there were so many that I had to ask the waiter to bring a fresh paper covering, folding the first one and tucking it away. In fact, we did have lunch in a restaurant a few days later, and we did sit at a small square table covered with paper. I smiled to myself as I made notes in a notebook. Over the course of several weeks and many months, we lunched often at the same restaurant. My list continued to expand as I learned more about science. At the very end of the project, Gerry and I met again. I had neglected to bring my notebook, the one in which I'd jotted down particulars about the scientists: what contributions they made and where I could find them. I searched for a piece of paper in my bag, to no avail. And the next thing I knew, Gerry said, "Just use the tablecloth!" It really was a dream come true.

The purpose of *Faces of Science* is to humanize scientists in a way not done before. In their texts, which they wrote or which were edited from interviews I conducted with them, I asked them to speak of themselves. What kind of childhoods did they have? How did they first become interested in science? What psychological needs may have drawn them to and been satisfied by their work? Why did they choose their particular field of research? Gerard Piel provided me with a specific "scientific" question for each subject to address. In response, some concentrated more on the scientific query, others on the personal. There is often a combination of the two. The portraits together with the texts offer a look into the humanity of an extraordinary group of people. Their intellectual curiosity, wish to help mankind, and ability to work with others to accomplish their tasks serve as a particular confluence of remarkable human qualities.

The completion of this collection of pictures prompts me to question whether one can generalize about the minds of scientists as distinct from the minds of other groups of people. While no generalization can be upheld exclusively, there is one specific trait shared by all the scientists I photographed. Each one is capable of highly sophisticated abstract thought and, at the same time, each has proven his or her conceptual hypotheses by practical experiment. It is the connection between these two very different capacities that requires an elasticity of mind uncommon to most of us.

The more scientists I photograph, the more additional ones I would like to photograph. It is wonderful to work with them. They are incisive, clear-thinking, and determined, and they manage their time better than any other "group" of individuals I know. It was very difficult to decide when to end the project. Finally, I took the historian Barbara Tuchman's advice concerning research to heart: "One must stop before one has finished. Otherwise, one will never stop and never finish." There are many scientists who ought to be in this book and are not. I hope I will have the privilege of photographing them in the future.

Introduction

GERARD PIEL

(Former publisher of *Scientific American* magazine)

THE PEOPLE TO BE ENCOUNTERED IN THIS BOOK ALL ENJOY the great good fortune of engagement in work that invites their utmost capacity. They are at work each on his or her carefully framed question about the nature of the physical world and of the life that wells out of it in our corner of that world. The work of the elders among them has had public verification in technologies that have changed the world we live in. That is what is expected of science: the technologies that make life richer and easier secure its public support. People who work at science find more immediate satisfaction. It is one that more of their contemporaries should share: what they are learning makes life in the 21st century a special privilege and the world around us still richer in the wonder we experience with our unaided senses.

How do stars make starlight? Here you will encounter Hans Bethe, who answered that question 70 years ago. He showed how, at the extreme temperatures and pressures in the depths of the Sun, for example, two atoms of hydrogen fuse into an atom of helium. The difference in mass between two hydrogens and one helium radiates as starlight, in accordance with the familiar equation $E=mc^2$. He will tell you more about his life and work. He also has something to say here about the public demonstration of his answer in the H-bomb and about the work he has done, with other scientists and citizens, to promote wiser use of the enormous forces that science has placed in human hands.

The stars visible to the naked eye and in the Milky Way galaxy we see arching across the night sky were once thought to be the whole universe. Working with the magnificent instruments on the mountaintops of California, Chile, and Hawaii, Allan Sandage, Maarten Schmidt, and their colleagues have carried human perception 10 orders of magnitude out into the universe beyond that little world. Each order of magnitude is a 10-fold increase; 10 of them is a 10-billion-fold increase in the radius of the observed universe. It can be seen that our galaxy is but one of billions of galaxies that throng in clusters and hang in curtains across the universal void. Cosmology, once the realm of theoretical speculation about the origin and history of the universe, is now an observational science.

The reach outward in space is equally a reach backward in time. That is because starlight travels at the constant velocity of 300,000 kilometers per second. Traveling as much as 13 billion light-years from the most distant galaxies observed, the light reports, as ongoing events, what was going on in those galaxies so far away and long ago. The scientists tell us that this was a stormy period in the then very much smaller universe. Martin Rees has shown that those galaxies are visible because they are consuming their substance, transforming matter to energy in the most brilliant light known to observers of the universe. They radiate on radio frequencies as well. David Helfand is mapping, in those frequencies, the distribution of these and nearer galaxies across the universe.

Radio frequencies carry the observed universe still deeper into the past. They have revealed a background radiation that comes from everywhere in the sky at a wavelength corresponding to a temperature of 2.725 degrees on the absolute scale, down near absolute zero. This is the first light that flooded the universe, cooled down, and stretched to longer wavelengths over the billions of years of travel into our time. The observed universe was a fireball at that moment, at the end of the first second of its history. Its radius was only one-millionth of its present radius, its temperature at 3,000 degrees.

Satellite measurement has shown that this universal background radiation does not vary by more than one part in five billion over the whole sphere of the sky. Alan Guth and Andrei Linde take this invariance to be evidence that the Big Bang—which started the expansion of the universe—was accompanied by a near-instantaneous inflation of the universe. That inflation, their calculations show, stretched the radius of the universe to 10,000 times the radius of the observed universe. It would appear that the observed universe is as tiny a relative to the whole universe as the universe in observation as late as 1920 compares with the now observed universe. We are not back where we started. But we must consider the possibility that what we have been calling "the universe" is, like our galaxy, just one of billions of universes, perhaps forever beyond our observation. This possibility is subject to further study of the background radiation, and calls on a grand novelty, foreseen by Einstein, that gravity also has a repulsive counterpart force.

Physics—now making its contribution to cosmology—has arrived at an almost complete understanding of the nearer world around us. Freeman Dyson put the finishing touch on the theory of quantum electrodynamics. This theory is the triumph of the international collaboration of two generations of humankind's best minds, starting from the work of Albert Einstein and Max Planck at the turn of the 20th century. Planck's "quantum" explained why ultraviolet light, for example, packs more energy than light with its longer wavelengths. Einstein showed that the quantum acts with the mechanical force carried by a particle, the mass-less "photon" of light. A sunburn demonstrates that force. Quantum electrodynamics has an appropriate acronym: QED. It stands as the final theory of electromagnetism, the force that illuminates the sky, holds atoms together, animates living organisms, and governs their environment here on Earth.

Wolfgang Panofsky and Leon Lederman helped to build two of the world's most powerful particle accelerators—"atom smashers." These giant instruments summon the electromagnetic force at energies that light a small city. In the inferno at their targets, particles accelerated close to the speed of light probe the particles in the nucleus of the atom—the proton and neutron—and discern the particles of those particles. These instruments carry human perception 12 orders of magnitude, a trillion times, below the tiniest thing visible to the naked eye.

At the outset, modest editions of these instruments challenged understanding with the discovery, in the structure of the atom, of two new forces of nature. Unlike gravity and electromagnetism, which act across the observed universe, these are short-range forces, acting within the dimensions of the atom and its particles. One is the strong force that holds the atomic nucleus and its particles together. The other is the weak force, in evidence in the breakdown of the atom and its particles. The energy of each of these forces has its quantum.

The finding of a multiplicity of forces was a kind of setback to the hope of the mission of science. In the words of Albert Einstein, science proceeds in the conviction "that the totality of all sensory experience can be comprehended [in] a conceptual system built on premises of great simplicity." There has since been progress toward that goal in the unification of the quantum forces.

Steven Weinberg secured in theory the unity of the weak force and electromagnetism—at an energy that pervaded the universe in its early history. That energy could be attained in accelerators, and the theory has been confirmed by experiment. Sheldon Glashow has essayed to bring those two forces into unity with the strong force, a unity calling for still higher energy at an earlier moment in the history of the universe. That energy exceeds by 10 or more orders of magnitude the energy of the most powerful accelerators. The Grand Unified Theory, or GUT, is nonetheless supported so far by what else is known from experiment.

Gravity, the first force of nature known to humankind, currently resists unification with the three quantum forces. That unification occurs, in theory, at energy levels still further beyond the reach of technology. Its proposed quantum, the graviton, has so far escaped detection by any instrument.

Failed by technology, the younger generation of physicists—John Schwarz is one of them—continues bravely with the building of theory. The "super-symmetrical string theory," on which he and others are working, holds the best promise of a proposal that can be confronted by experiment.

The accelerators have, meanwhile, brought resolution of the structure of the atom. From the explosions of short-lived particles at the machines' targets, Murray Gell-Mann has shown that the neutron and the proton must be composed of different combinations of what, borrowing from James Joyce, he has called quarks. He has shown, further, that the strong force—here organizing the quarks into protons and neutrons—is carried by three different quanta. He has named them, poetically enough, after the three primary colors that, as shown by Isaac Newton, combine to make white light. In accordance with that metaphor, his comprehensive theory goes by the name of quantum chromodynamics. As QCD, it stands alongside QED.

Symmetry, such as that found on the two sides of an equation, is as central to the physical theory as it and its violation are to aesthetics. That is why the community was rocked by the proposal of Chen Ning Yang that symmetry would be violated in the decay of a certain class of particles. The prediction was confirmed in accelerator experiments conducted by Leon Lederman. The work of Val Fitch resolved this quandary. It explained why the universe is made of matter, and not the antimatter that turns up in the torrent of particles from the accelerator targets. Such antimatter mutually annihilates with matter on the instant of its appearance. The observed universe owes its very existence, it turns out, to the asymmetry of matter-antimatter. The matter of which the universe is made is the tiny excess of matter left over from the matter-antimatter annihilation in the Big Bang.

Here in our warm corner of the universe, the tendency of matter to self-organize—two atoms of hydrogen and one of oxygen make a molecule of H_2O, water—has arrived at super-organization in living organisms. To us living organisms, there is no deeper mystery than life. The 20th century brought a breakthrough to a start on understanding. Early in the century, three botanists, independently and simultaneously, made the same discovery: sexual reproduction conveys a genetic trait from one parent or the other—never a blend or mixture of the trait in the two parents. Then they made a second simultaneous and independent discovery: another botanist, Gregor Mendel, had made that discovery nearly half a century before. Mendel had called whatever it is that conveys a trait the "gene."

The rediscovery of the gene set many scientists to determining its nature. They searched, especially, among the proteins that are the principal constituents of the living cell. Then, in the 1930s, Oswald Avery, Colin MacLeod, both deceased, and Maclyn McCarty, present here, showed that a quite different molecule, a rarer constituent of the cell, does the work of the gene. The genetic molecule, as everyone knows today, is DNA, or, as fewer people can say, deoxyribonucleic acid.

Resolution of the structure of DNA now became the grail of inquiry into life. Francis Crick and James Watson accomplished that task in 1953. They showed that DNA functions as a double helix: two long strands of DNA wound helically around each other. One helix encodes the genes, the instructions for the fashioning of each living organism from other molecules. This strand does the encoding in successive clusters of three of the four submolecules—called bases—that compose DNA. The second strand arrays the DNA bases in complementary reverse sequence. In reproduction of the genes, this strand attracts the DNA bases and fixes them in a new strand in the gene-encoding sequence.

The very first question asked by this discovery was "How does the complementary strand, bound to its partner, get loose to conduct that reproduction?" It was answered by what has been called "the most perfect experiment in biology," performed by Matthew Meselson, present here, and his colleague Franklin Stahl.

Next in the host of questions now opened up to inquiry was "How does DNA convey the instructions it encodes for expression of genetic traits in the synthesis of protein molecules?" The answer was found in RNA and its almost identical set of bases. Assembled on the complementary strand of DNA as shown by Walter Gilbert, RNA governs the assembly of the protein molecules from their subunits. Each RNA base corresponds to a subunit of the encoded protein molecule. Among these proteins are the regulatory "repressors" that turn the genes on and off, as shown by the work of François Jacob and his now deceased colleagues André Lwoff and Jacques Monod.

Life begins to be understood as the manifestation of an ordered web of interlocked chemical reactions that start and stop one another. The dawn of such understanding has enlisted molecular biologists in ambitious enterprises. A half dozen of the scientists here are at work on the molecular anatomy and physiology of the living cell. The genesis and organization of cells in the embryonic development of multicelled organisms engage Robert Edwards and Nicole Le Douarin. The international "worm group," inspired by Sydney Brenner, aims at the total molecular biological description of a model organism, the nematode *Caenorhabditis elegans*. Norton Zinder has achieved such a description of a 10-gene virus of the sort that infects bacteria. In the simple nervous system of another model organism, the marine snail *Aplysia*, Eric Kandel has demonstrated the basics of the molecular biology of memory and learning.

Successful laboratory experiments provide the technology for the next round of inquiry. The highly mechanized technology that recognizes the DNA bases has now sequenced the bases in the genomes—the complete array of genes—of the dozen or more animals and plants that serve as model organisms. Eric Lander led the international consortium of university laboratories that sequenced the billions of bases of the human genome. It remains to identify the full complement of genes encoded in the bases. To the dismay of some people, it turns out that the same number of genes—on the order of 30,000—that blueprints other vertebrates also specifies the human being. Most of the genes are identical; in the case of other primates, almost all are identical.

The long history of evolution recorded in the human genome somewhat complicates the molecular biology of human tissue cells. The chains of DNA encode nonsense over long stretches between genes. The nonsense, apparently, are the fossils of viral infections that beset our ancestors and the primate line before them. Neutralized in the survivors, the virus genes were incorporated in the DNA chains. James Darnell showed that the RNA "edits" the nonsense out of its chains before it engages in its vital functions.

It begins to be possible to trace evolution back to its very beginning and ask how life got started on Earth. On evidence from geology and even the finding of fossil bacterial cells in the rocks, William Schopf and other interested geologists have carried the origin of life back 3.8 billion years. That would be just a few hundred million years after gravitational collapse formed the planet and soon after it cooled down enough to hold water in the liquid state. The finding of traces of water on Mars would suggest that life got started there as well.

Many scientists are looking for the self-replicating molecule that began the story. Leslie Orgel staked his career on RNA and what might have been its precursors. He shares the frustration of those many colleagues. The discovery by Stanley Prusiner of the prion—a self-replicating protein molecule implicated in mad cow and Alzheimer's disease—has put proteins in the race.

Life got started and persisted here because Earth, unlike Mars, remains geologically alive. The incandescent heat of its formation still keeps the nickel-iron in its core in the liquid state. Lynn Sykes and many colleagues have shown that immense horizontal forces originating in the interior have constantly remade the map of the world over its long lifetime, opening and closing oceans, assembling and breaking up continents. The history of life is necessarily much engaged in the history of the planet. Paul Hoffman has established a time, 700 million years ago, when global warming triggered the onset of an ice age that froze the entire planet in a "snowball," with glaciers grounded on the sea bottom at the edge of continents in equatorial latitudes. That global warming was brought on largely by the runaway success of the single-celled bacteria that teemed in the world ocean. Their respiration increased the concentration of carbon dioxide, the principal "greenhouse" gas, in the atmosphere, much as the burning of fossil fuels does today. Global warming slows down and stops the great ocean currents, like the Gulf Stream, which keeps Europe warm—London, at close to 51 degrees latitude, is 700 miles north of New York (latitude near 41 degrees).

Steven Stanley and others have found that the snowball Earth played a critical role in the origin of multicelled organisms. The freeze-up reduced the population of bacteria long enough to make room for them. Their barely discernible fossils turn up first in mountains upheaved in what is now Arctic Canada; in no time at all geologically speaking, the fossils of their descendants imprinted the rock all around the world.

These organisms were made of a new kind of cell, the eukaryote. It had originated in the ocean, perhaps 500 million years earlier. Its larger genetic capacity brought the segregation of its DNA in a nucleus within the cell, instead of floating free in the protoplasm as it does in the primordial bacteria. Lynn Margulis has shown that the eukaryote arrived by an evolutionary shortcut. The different species of bacteria, which had by then evolved through the slow process of genetic mutation, formed mutually sustaining colonies. Close symbiotic interdependence brought the fusion of bacteria of different kinds in a single cell. The organelles of our tissue cells, especially the oxygen-consuming, energy-generating mitochondria, give evidence of such fusion. In the green leaf, another organelle conducts photosynthesis. Further, Margulis argues, symbiotic association of eukaryotes precipitated their closer association in the first multicelled organisms.

By 400 million years ago, those simple organisms had differentiated into ancestral models of the existing phyla of plants, animals, and the humble fungi. The phylum is the senior level in classification of the diversity of genera and species that has reached its all-time peak in our time—and human occupation of the planet has set extinction going at rates recorded only in planetary disasters, such as the impact of the massive meteorite thought to have carried off the dinosaurs. In quick order the phyla, as shown by the late George Gaylord Simpson, diversified—in his vivid term "radiated"—into genera, and the genera radiated into species. Francisco Ayala shows how new species arise in large populations of animals that proliferate into different environments and come under differing pressures of natural selection.

Richard Leakey carries on the work, so well started by his parents, Louis and Mary Leakey, into the origin of man. In 1959, the senior Leakeys discovered the bones of the first toolmakers with their tools in Olduvai Gorge, in Kenya. Careful dating carries that site back 1.5 million years into the past. The bones were the bones not of human beings but of a 90-pound primate. Not far away, at Lake Turkana, they found toolmaking "workshops," with raw material carried in from distant "quarries." They found also—by this time Richard was working with them—the jawbones and incomplete skulls of 100 individual primates. They made confident reconstruction of six skulls. The skulls of the toolmakers, whom they named *Homo habilis,* were scarcely distinguishable from those of the non-toolmakers, their cousins. Toolmaking has now brought the later arrival *Homo sapiens* into custody of the planet, exceeding by 100,000 times the number of other animals of comparable size—except for his pets and farm animals.

For all the light that the past casts upon the present, life and the world around us press an infinity of questions to be asked and answered. The question of vision has carried the tracing of nerve circuitry deepest into the brain. The late Stephen Kuffler and his colleagues David Hubel and Torsten Wiesel, present here, showed that the first processing of the image on the retina occurs in four layers of circuitry just behind the retina. They have followed the translation of sensation to perception through further circuits, deep in the visual centers of the brain. Maize—the New World's gift to human nutrition and now the world's third grain crop—engages the interest of Mary Eubanks and Nina Fedoroff. Eubanks has settled the vexed question of the ancestry of corn, showing that hybridization of two grasses, *Tripsacum* and teosinte, brought the separation of the tassels, the male organ, from the ear and its silk, the female organ. Fedoroff carries forward the discovery by Barbara McClintock of "jumping genes," developing the importance of these "transponsons" in the unstable genetics of this important plant. Ruth Patrick and Miriam Rothschild are naturalists. Patrick is the founder of "limnology," the ecology of pond and brook. Rothschild, tirelessly intrigued by new questions, showed how the flea makes its high jump: its metabolism stores energy in its "knee" tendons, expended in the instant of the jump. A great many more women will appear in any such future assembly of scientists.

With all the interesting people to be encountered here, this introduction must stop. The work of the scientists, especially that of the younger ones for which this introduction is too short, will carry the same surprises for and extension of human understanding—and some of it will inevitably find its public verification in technology. Nor do space and time allow consideration of the often ardent controversy and the competitive collaboration with which the community of scientists carries the work forward. That work compels these human beings, however, to exercise the faculties of reason, tolerance, and mutual respect. That is the sort of community in which the rest of us would be happy to live.

Faces of Science

Francis Crick

Biophysics. Nobel Prize in Physiology or Medicine, 1962, for discoveries concerning the molecular structure of nucleic acids and its significance for information transfer in living material.
J. W. Kieckhefer Distinguished Research Professor, Salk Institute for Biological Studies.

Here in a distant place he holds his tongue,
Who once said all his say, when he was young!

— James Stephens, "Egan O Rahilly"

Richard Axel

Representation of sensory information in the brain. Nobel Prize in Physiology or Medicine, 2004, for discoveries of odorant receptors and the organization of the olfactory system. University Professor of Biochemistry and Molecular Biophysics and of Pathology, Columbia University College of Physicians and Surgeons. Investigator, Howard Hughes Medical Institute, Rockefeller University.

New York City is my world. As the son of immigrant parents with a deep respect for learning, I moved from public schools in Brooklyn to high school in Manhattan, and a new culture was revealed. I was exposed to art, books, music, and science for the first time. I entered Columbia College at a time of discord, the early sixties, uncertain of my path. Would I become a physician following the wishes of my parents, a scientist, or pursue my interest in literature? With the demonstration that DNA is the repository of genetic information, with the elucidation of the structure of DNA and the deciphering of the genetic code, it was clear that I wanted to play a part in the "new biology."

My plans were thwarted by an unfortunate war. To assure a deferment from the military, I found myself a displaced medical student at Johns Hopkins. My clinical incompetence was immediately recognized by the faculty and deans, and I was allowed to graduate with the understanding that I never practice medicine on live patients. This left me little alternative but to pursue a career in molecular biology in earnest. I entered biology with the good fortune of participating in a revolution made possible by recombinant DNA technology. We learned how to cut and paste DNA, to isolate the genes, and to manipulate them at will. In my laboratory, we recognized that the ability to isolate genes would require assays to understand gene function. Within months after establishing my laboratory at Columbia, we devised procedures that permitted the introduction of virtually any gene into any cell, a gene transfer system that allowed not only for the isolation and expression of genes but for a detailed analysis of gene function.

In the early 1980s, I became scientifically restless, and the rich neuroscience community at Columbia began to teach me about the brain. It became clear that the brain of different organisms must be endowed with an a priori potential to perceive the world and to respond to it. How we perceive the external world and how sensory stimuli elicit thoughts and behaviors must, in part, be made possible by our unique allotment of genes. This wholly reductionist stance argued that molecular biology could provide an understanding of how our sense organs combine with our brain to perceive the world. This excited me, and my laboratory, previously dedicated to molecular biology and genetics, now began to interface with neuroscience to approach the tenuous relationship between genes, behavior, and perception.

Most recently, my laboratory has been interested in the elusive sense of smell, how olfactory information in the external world is represented in the brain and how this representation can lead to meaningful neural information, thoughts, and behavior. We have isolated 1,000 genes that are involved in the discrimination of odors, the largest family of genes in the chromosome. The understanding of how these genes function has afforded surprising insight into how genes shape our perception of the sensory environment. Science for me is a joyous obsession, and I look forward to the continued excitement of future experiments in the perception of the sensory world amid the cultural fire of New York City.

Francisco J. Ayala

Population and evolutionary genetics.
Donald Bren Professor of Biological Sciences and University Professor of Philosophy, University of California, Irvine.

The first science class that I remember was in junior high, in Madrid, where I was born. The Catholic priest who taught the class got me hooked. I wanted to learn about astronomy, physics, and biology, about scientists and their exciting discoveries. When I registered to study physics at the University of Madrid, my parents, brothers, and sisters were surprised and a bit chagrined. Nobody in the family had ever studied science. Their interests were in business; economics or law would have been expected, but not physics.

While at the university, I read *Le Phénomène humain* (The Phenomenon of Man), by Pierre Teilhard de Chardin, the Jesuit anthropologist and philosopher. The book presented a vision of an evolving world of matter and life that has so far culminated in the human species, with greater developments to come. I would later consider this book more a poetic manifest than a scientific treatise, but at the time it inspired me to study evolution and genetics.

Around 1960, science in Franco's Spain was in a sorry state. Two Spanish professors whom I had befriended encouraged me to go abroad and introduced me to Theodosius Dobzhansky, one of the great evolutionists of the 20th century. And so it was that in 1961 I began doctoral studies at Columbia University with Dobzhansky as sponsor. I was blinded by the excitement of New York City. I would study and work at the lab all day, including most weekends, but I would often attend poetry readings or music performances in the evening, mostly in Greenwich Village, and go to see foreign and experimental movies that had never been accessible in Spain. At the university, the facilities were splendid and the professors were brilliant scientists, accessible and friendly, a far cry from the stuffiness that prevailed in Spanish universities at the time. The skyscrapers were not claustrophobic as I had assumed, but rather reached to the sky and framed the immensely long, straight avenues, which never ended but extended into the sky, at the horizon. The first Sunday I visited the Museum of Modern Art and was overtaken by emotion when I discovered Picasso's *Guernica*.

In 1964, I got my Ph.D. degree. In 1967, I got my first faculty position, at Rockefeller University in New York. Dobzhansky was doing as much as he could to persuade me to stay in the United States. In March 1971, I became a U.S. citizen and later that year I moved to the Department of Genetics at the University of California at Davis. In 1987, I moved to the Irvine campus of the University of California, where I shall remain to the end of my career.

Genetic diversity and change have been the subjects of my research. I have investigated the changes that yield new species, how to use DNA to learn about the evolution of organisms and their interrelationships through time. The millions of "letters" that make up the DNA of organisms are books where we can read the history of the species and the timing of evolutionary events. For the last two decades, I have invested much effort in unraveling the genetic traits of the malaria parasites that every year kill a million children in Africa alone and debilitate with high fever hundreds of millions of infected individuals throughout the world.

Scientific research is exciting. I wake up every morning eager to get to my office and lab to pursue the subject at hand with my graduate students, postdocs, and other collaborators. I very much enjoy teaching undergraduates. For ten years, I have taught introductory biology to 500 students each quarter. And I enjoy world travel: for field research, for lecturing, for vacation. The little friend on my lap was a gift from Indians living in the jungles of Colombia.

Neta Bahcall

Astronomy. Professor of Astrophysics, Princeton University.

Growing up in Israel, I fell in love with math and science in high school. I was inspired by my outstanding science teachers. They were fun and they were exciting. I loved the precise logic and the amazing ability of math and science to calculate, quantify, and predict most of the events occurring in nature: What are the atoms made of? What are we made of? What causes day and night, summer and winter? Are there other planets in the universe? How do we find a cure for cancer? I was excited by the questions, and I loved the logic. Once you understand the fundamental equations of nature, you feel you can solve almost any physical puzzle. I graduated in physics and math from the Hebrew University in Jerusalem. At that time, in the early 1960s, there was no astronomy in Israel.

There were no scientists in my family. My father was a lawyer. He wanted me to become a lawyer but I was much more attracted to science. My mother was a registered nurse. I thought of becoming a medical doctor. On my frequent visits to my mother's workplace in the hospital, I felt the importance and the satisfaction of helping sick people. I always felt that being a doctor is a noble profession.

I entered the field of astrophysics after I met my husband, John. We met at the Weizmann Institute, where he was visiting and I was studying for my graduate degree in physics. After we married and moved to Caltech, where John was on the physics faculty and I worked on my Ph.D. program, I opened up to the exciting field of astronomy. Caltech is one of the top places in the world for astronomical research. I met and talked with legendary astrophysicists—William Fowler, Maarten Schmidt, Fritz Zwicky, Geoff and Margaret Burbidge. This was love at first sight for me. I loved being part of an effort to answer big questions: What is the universe made of? How did it form? How big is the universe? What are the most distant objects? I still find it mind-boggling that we can ask and now answer such fundamental questions on such majestic topics.

My research involves some of these exciting questions: How much does the universe weigh? How much dark matter exists in the universe? Will the universe expand forever? These are among the most fundamental questions in astrophysics and cosmology, and they are so exciting—like falling in love again and again. Our group was among the first to show that the universe contains only about 20 percent of the critical mass density needed to halt the universal expansion. (The universe began expanding at the Big Bang, about 14 billion years ago. The mass in the universe, through its gravitational pull, slows down the expansion, just as the brakes slow down a speeding car. The heavier the universe, the more it slows down to stop the expansion.) We weighed the universe using several independent methods; all gave the same result: there is not enough mass to stop the universal expansion. This result was unexpected and was originally resisted. Our group was also among the first to find unexpectedly large-scale structure in the universe: galaxies and clusters of galaxies organized into huge superclusters—like large mountain chains on Earth—extending hundreds of millions light-years. Such enormously large structures would be expected only with a lightweight universe. This picture of a lightweight universe, with only about 20 percent of the critical density, has now become a "standard" model of the universe, confirmed by numerous recent observations. We are proud of being among the first to show this result.

Most of the inferred mass is dark—that is, it does not shine. We know its existence from its gravitational effect on other nearby objects. Most astronomers and physicists believe the dark matter is made up of some yet unknown elementary particle of physics. Many experimentalists are currently trying to detect and identify the dark-matter particle. The "normal" luminous matter in the universe, found in stars and gas, and made up of familiar atoms, is only about 5 percent of the critical density.

The current "standard" model of the universe also suggests that, in addition to dark matter and normal matter, a mysterious dark energy exists. The dark energy, which exists in empty space and opposes the pull of gravity, causes the universe to expand faster and faster as time goes on. (Dark energy was discovered recently when it was found that the expansion of the universe, quite surprisingly, is accelerating rather than slowing down as previously expected.) It will eventually become much larger, much darker, and much colder. But this will happen only in many, many billions of years.

I am frequently asked, "Do these studies of the vast universe make you feel that we are all small and insignificant?" My answer is no, not at all; in fact I feel exactly the opposite. I feel that we human beings are incredibly smart to be able to sit on this small planet of ours and figure out such amazing things about the universe. I find it to be enormously empowering. I always share this feeling with my students and with the public. And our scientific insights teach us how beautiful and fragile and precious our planet Earth is, and how much we need to take care of it—all of us.

David Baltimore

Molecular biology. Nobel Prize in Physiology or Medicine, 1975, for discoveries concerning the interaction between tumor viruses and the genetic material of the cell. President, California Institute of Technology.

Looking back on over 40 years in science, I am amazed by the variety of activities and issues in which I have been involved. Amazed because I started out on this odyssey with a simple step, the decision to go to graduate school and then to study viruses. That decision unwittingly but quickly made me a pioneer in the then developing field of molecular biology. And when, 10 years after deciding to devote my experimental life to viruses, I discovered from a virus that RNA can be copied into DNA—providing a counterpoint to the central dogma that DNA makes RNA and winning me a Nobel Prize 5 years later—it all seemed like a natural extension of the original decision. From there to building and maintaining scientific organizations was a large step, taken slowly and developed in counterpoint to interests in immunology and neuroscience.

What have I learned in all this life? Mainly to trust my instincts and follow what seems to give me satisfaction. And not to push too hard professionally. I've left the pushing to the recreational side of my life, where I keep trying to do things at which I know I will never excel but from which I get moments of very personal pleasure: skiing, fishing (especially), sailing, even windsurfing. Now add that to music, art, eating, drinking, and owning too much wine, and the picture is virtually complete. The one other crucial piece is the people I love and the others I know and care about.

The discordant piece of my life for many people might be the running of institutions. Scientists are supposed to be wooly-headed nerds who eschew dealing with real-world issues. The pull on me to influence institutional governance dates back to my earliest days in science when I was at a new institution, the Salk Institute, and the importance of governance was made evident to me. When I had the rare good fortune to be asked to start a new institute myself, I saw it as an opportunity to put into practice my thinking about how the ideal institution would work. Thus, I started the Whitehead Institute, now a premier venue for biomedical research. That led to my being president of the California Institute of Technology, Caltech, where I currently reside.

Science in America today has seemingly lost its luster while at the same time being the driving force of the economic strength of the country and the engine that leads to an increasingly rich and varied life for all Americans. It is also the source of our security as a nation. This contradiction cannot be maintained if we are to remain the world's leading nation. We need to find ways to return science and engineering to the pedestal it had during the post-Sputnik era. This is an important role for institutional leaders, but we alone are unable to have much influence. The impetus has to come from parents and schools because the decision to be a scientist has to be made early. For me the final decision was in high school, but even earlier I had emphasized a curriculum that would support a life in science.

Cori Bargmann

Brain studies. Neuroscience. Torsten N. Wiesel Professor and Head of Laboratory of Neural Circuits and Behavior.
Investigator, Howard Hughes Medical Institute, Rockefeller University.

The human brain is a wonder of nature, the seat of our thoughts, memories, desires, and emotions. But the brain is also a biological organ, and like all biological organs, it is built by genes. No serious person believes that genes control our behavior in a deterministic way, but no serious person denies the importance of genes either.

Any individual person's or animal's behavior arises from a number of elements: a genetic component, the environmental context, and the history of the individual's behaviors and experiences. Of these elements, the genes are the easiest part to understand, and that's why we study them. There are infinitely many environments and experiences, but only a certain number of genes.

We, my students and postdocs and I, try to understand how genes affect the brain by studying the genes and behavior of the microscopic nematode worm *C. elegans*. Humans are much more complicated than simple animals like worms and flies, but we share many of our genes—about half of all genes in humans are present in these simple animals. The degree of similarity came as a surprise even to those of us who love the worm. For the genes that are shared, we can study the function of a certain gene in the worm and learn something about what that gene does in all other animals, including ourselves.

Behavior is complex, unpredictable, chaotic, even in simple animals. We don't always know what question we're trying to answer when we start. We explore an area that we don't understand, and then try to describe what's important or interesting about it.

Most of our studies have centered on the worm's sense of smell. These worms don't see or hear, so smell is their main way of gathering information about the world. They're very good at it, because evolution has selected for their success. Genetically, their sense of smell is more complicated than ours.

We found a gene that was required for the worm to smell one particular odor, the smell of buttered popcorn. The gene encodes the receptor for the buttery odor. The receptor is a molecule that sits on the surface of a nerve cell in the nose. Part of the receptor is outside of the nerve cell, and part is inside the nerve cell. The receptor is like a TV antenna on a house, waiting to detect signals from the outside world. When the right odor comes along, it binds to the receptor, which then sends a signal to the nerve cell to tell the inside of the cell which odor is present. The nerve cell converts this information into electricity— the language of the nerve cells—and sends it to the brain.

Normal worms love the buttery smell, but if we move the receptor into a different nerve cell, a nerve cell that detects danger, the worms detest the buttery smell and avoid it. This experiment tells us that nerve cells in the nose are prewired into behaviors, either attraction or avoidance. The nerve cells in the nose encode instinctive preferences, which are widespread in nature. Human newborns accept sweet flavors, and reject bitter flavors. Since many bitter compounds are toxic, this instinctive preference works to our advantage.

We found a gene that creates differences between the behavior of individuals. Some worms are solitary, and other worms gather into social feeding groups. Both behaviors are widespread in nature, and both kinds of worms seem to coexist in the same environments. The difference between solitary and social worms arises from a single gene, and from a single change in that gene, everywhere in the world. I don't think that this exact gene is the reason that I'm shy, but I think we may be getting at the basic pathways that regulate behaviors.

Paul Berg

Biochemistry. Molecular Biologist. Created the first "recombinant DNA" molecules in 1972, thereby creating the field of genetic engineering. Nobel Prize in Chemistry, 1980, for his fundamental studies of the biochemistry of nucleic acids, with particular regard to recombinant DNA. Robert W. and Vivian K. Cahill Professor of Cancer Research, Emeritus, Stanford University.

Discovering something that no one had ever known before can be a very heady experience, especially to a high school science student. I learned that early in a high school science club where an inspiring science "teacher" pushed me to learn by doing, by asking questions and devising experiments to answer those questions. Whether fictional (as in *Arrowsmith*, by Sinclair Lewis) or biographical (like the medical pioneers in *Microbe Hunters*, by Paul De Kruif), scientists became my heroes. But I was drawn to biochemistry rather than medicine because discovering the metabolic machinery of humans seemed more challenging intellectually. A hitch in the navy during World War II was only a "bump in the road," and soon thereafter I completed an undergraduate degree and a Ph.D. in biochemistry.

After receiving the doctorate, I studied a year in Copenhagen at the university's Institute of Cytophysiology, after which I joined the faculty at Washington University in St. Louis. There, triggered by the discovery of the DNA double helix and the ensuing creation of molecular biology, I made the slow transition from classical biochemistry to molecular biology. After moving to Stanford University, I became preoccupied with how genes act and proteins are produced.

In the late 1960s, my interests shifted to how SV40, a small virus, causes cancer in rodents. It was known that the virus's ability to insert its DNA genome into the chromosomes of cells it infects was responsible for its carcinogenic property. Because of its ability to transform the genetic makeup of cells it infects, I considered that it was possible that SV40 could be used as a carrier to introduce foreign genes into mammalian cells. Following up on that idea, I and two colleagues devised a general procedure to "sew" bacterial genes into the SV40 DNA chromosome. That was soon followed by a more efficient procedure to clone genes, indeed any DNA segments from any organism, living or dead, on our planet. Cloned and amplified genes were valuable objects for basic structure studies. As importantly, the cloning technology made it possible to derive the entire genetic structure of human and hundreds of additional related and unrelated organisms. Recombinant DNA or, as the public prefers, gene cloning, or genetic engineering, provided the means to produce commercial quantities of therapeutic proteins like insulin, growth hormone, and blood-clotting proteins, and the biotechnology industry was born.

The use of the recombinant DNA technology now dominates research in biology. It has altered both the way questions are formulated and the way solutions are sought. The isolation of genes from any organism is now routine. Equally profound is the influence it has had in many related fields. Even a brief look at journals in such diverse fields as chemistry, evolutionary biology, paleontology, anthropology, linguistics, psychology, medicine, plant science, and, surprisingly enough, forensics, information theory, and computer science shows the pervasive influence of this new paradigm.

But the most profound consequence of the recombinant DNA technology has been our increased knowledge of fundamental life processes. No longer is the gene an abstract notion, nor is it as enigmatic as interstellar dark matter or black holes. Genes, and chromosomes of which they are a part, are describable in precise chemical terms. Even more significantly, genes can be synthesized in test tubes, manipulated, and reintroduced into the cells of living organisms, enabling us to link genes with specific physiological functions.

The public, however, was less enthusiastic than the scientific community. During the middle and late 1970s, the debate raged over gene cloning and the concerns about what genetic engineering might bring. The debate was heated, divisive, and focused on the most unlikely catastrophic, even apocalyptic outcomes. For the first time, legitimate biomedical research faced threats of federal prohibition. Largely through the lobbying efforts of scientists, physicians, and the budding biotechnology industry, the attempts to prohibit recombinant DNA experimentation in the United States were blocked. Instead, research and development proceeded according to guidelines that mandated oversight by the National Institutes of Health. Lacking any congressional action, the science flourished, commercial opportunities were realized, and the United States now leads the world in the developing biotech industry.

Hans Bethe

Theoretical physics. Nobel Prize in Physics, 1967, for his contributions to the theory of nuclear reactions, especially his discoveries concerning the energy production in stars. John Wendell Anderson Professor of Physics, Emeritus, Cornell University.

Born in 1906, I look back on most of the 20th century. It was a century that saw many important developments in science. Quantum theory was the most important of them, followed by the discovery of DNA and then by Einstein's theory of special relativity. Quantum theory gives you the key to all atomic phenomena, and that in turn gives you the key to chemistry and biology. DNA is fundamental to biology, and special relativity is essential for exploring nuclear physics.

There were also many important inventions during the 20th century. The three most important ones, I think, are the airplane, the transistor, and the computer. The airplane gives us communication with the whole world. The transistor is fundamental to all electronic communication. What the computer has done to affect our lives hardly needs elaboration.

My birthplace was Strasbourg, in what was then part of Germany. I grew up in Germany, was educated there, and began my career in research and teaching there. But, like so many scientists of Jewish ancestry, I left Germany with the advent of the Hitler regime. After two years in England, I received an appointment at Cornell, with which I have been associated ever since.

At Cornell I have worked on many aspects of physics, but my interest in the source of the energy in the Sun and larger stars probably drew the most notice. It was clear in 1938 that the energy emitted by the Sun and many other stars was due to nuclear fusion. In a fusion reaction, two light nuclei collide and fuse, forming a new nucleus. There is a loss of mass in this reaction, the mass being converted into radiant energy. It was not clear at the time what was fusing into what. The basic reaction was thought to be the fusion of two hydrogen protons, producing a deuteron. That reaction does not liberate much energy, but the deuteron starts a chain of reactions, the net result of which is that hydrogen is transformed into helium, with the release of a great deal of energy. Calculations that I and others made of the energy released agreed with the actual rate of energy production in the Sun.

It did not seem, however, that these findings explained what was going on in bigger stars. I went through the periodic table of the elements, looking for nuclei that might be involved in fusion reactions in such stars. Nothing seemed to work. But when I tried carbon, it worked. A carbon cycle is the source of energy in larger stars. I see this as a discovery by persistence rather than by brains.

In a separate but related career, I have been deeply involved with the issue of nuclear weapons. I was involved in developing the atomic bomb during World War II because I did not want to see the Nazi regime succeed at that endeavor first. I have no regrets about that work. But after the war, appalled by the devastation wreaked by the atomic bombing of Japan, I spent much of my time urging the United States and scientists everywhere to renounce research on nuclear arms. As a member of the President's Scientific Advisory Committee, I pressed for what became the world's first and most successful arms control pact—the Limited Test Ban Treaty of 1963, which required future nuclear tests to be underground. I can sum up my feelings on this subject by citing what I wrote to President Clinton in 1997, urging a ban on "all physical experiments, no matter how small their yield, whose primary purpose is to design new types of nuclear weapons."

Elizabeth Blackburn

Cell biology. Cancer Research. Professor, Department of Biochemistry and Biophysics, University of California, San Francisco.

I always loved animals, and I was fascinated by anything living. When I was a teenager, I learned about biochemistry and proteins, and I got the feeling that the answer to understanding biology would lie in understanding what went on in cells, and how you could understand that at the level of molecules. I was very excited about that, but it really grew from a love of animals and of the natural world. There's a kind of pantheism in me which takes the form of wanting to understand the world and feeling impressed and awed by the natural world, but particularly life. My parents and many of my family were, and are, physicians; I think there's a physician gene in my family. But while my older sister became a physician, I was the rebel and decided to be a scientist. Of late, though, I've become more interested in how our science takes us back to questions of health and cancer.

The science I do is all about telomeres. The telomere is the name given to the end of a chromosome. There's a scientific story there, and there's also a personal side. For many women in science, Barbara McClintock, now deceased, is a heroine. I was greatly impressed by her when I talked with her on a number of occasions. I had heard of her famous work on jumping genes. But she was also the first to discern that the natural end of a chromosome is different from a broken end of a chromosome. Unlike an accidentally broken end, it's very well protected against ever fusing with anything else. If it does try to fuse, that causes genetic instability, which is bad news. Later, the renowned fruit fly geneticist Hermann Muller gave the natural ending of the chromosome the name "telomere."

There is a good metaphor: if a shoelace lacks that little aglet on the end, the shoelace starts fraying and it also won't behave properly; you can't tie it properly, and it won't go through the lace hole. The telomere is like the aglet of the chromosome. If you don't have telomeres on both ends, the chromosome ends start fraying away. I was the first to look at that aglet DNA. I had worked in the 1970s with the person who discovered how to sequence DNA, Fred Sanger of the University of Cambridge. We loved DNA sequences then, and I wanted to sequence the end of the DNA of a chromosome. First, I worked with very short minichromosomes discovered by my postdoctoral fellowship adviser Joe Gall at Yale. We could get lots of ends, and I sequenced them. I discovered what the DNA sequence was like at the end of the chromosome: it was very weird and unexpected. Then later, from looking at various things happening to it, I thought there must be a novel enzyme that adds more DNA to telomeres, to keep replenishing the ends. Indeed, the ends are prevented from fraying because they get replenished by such an enzyme, which I discovered, together with my graduate student Carol Greider, at Berkeley. It was a completely new enzyme; no one knew it existed before. It was very exciting. We called it telomerase.

The way this relates to cancer is that most cells restrict the amount of multiplication they can do. The reason we are the shape and size we are is that our cells don't multiply too much. But a cancer cell loses controls, and it keeps multiplying. Then it's in danger of its ends fraying away unless telomerase comes and replenishes them. Telomerase gets turned up very high in cancer cells. As a result, a cancer cell's telomeres never fray away, and the cell can just keep multiplying.

There's a converse side to telomerase in people, though. To get you through your full normal life span, you need to inherit two good working copies of a particular telomerase gene, one from your mother and one from your father. If you received, from one parent, a version of that gene that does not work, even though the gene you got from the other parent works, that is not enough. Sadly, in people where this happens, the bone marrow becomes exhausted and they die in early adulthood or in middle age. Thus even half the normal amount of telomerase can't take you all the way through to old age.

What use can we make of all this in a medical school? In my lab, we'd like to cure cancer. I decided that the fact that cancer cells have turned their telomerase up full blast could be made into a vulnerability for cancer cells, and that we could make them regret that their telomerase is turned up. It's like jujitsu; you take the force of your opponent and turn it against him or her. Also, we recently discovered that cancer cells become addicted to their telomerase, and if you withdraw it, they will kill themselves, which surprised us: the cells know right away, and they commit suicide. We're really just learning about this addiction in the lab and in certain model tumors.

The best way of doing all this will be with small drug-like molecules. We all know it's a very long route before it will ever be of any use to a human being with cancer. Cancer gets cured every week but only in the lab, so I'm realistic. On the other hand, when you're a scientist, you have to be both realistic and optimistic. You may think of a thousand reasons why an experiment could fail, but you have to try it because it's important to use some intuition in your science and move forward with the idea.

Sydney Brenner

Molecular biology. Nobel Prize in Physiology or Medicine, 2002, for discoveries concerning genetic regulation of organ development and programmed cell death. Distinguished Research Professor, Salk Institute for Biological Studies.

Inside the cells of my body are hundreds of tiny molecular machines whose smooth and endless working ensures my existence, and drives the thoughts in my head and the words I write and speak. Their intricate structures are specified by my genes, which are the products of billions of years of evolution and trace a continuous lineage between me and our simple primordial ancestors. In the course of this immense period of time, many living organisms have risen, only to fail the stringent test of natural selection. They fell by the wayside, and you and I and every other contemporary living organism are the surviving strands of the unfolding past.

Understanding these marvels has been the adventure and the labor of my life. Beginning with the discovery of the structure of DNA, much light has been thrown on these intricate molecular processes by the past 50 years of research, but much remains mysterious. I wish I had a second life and could live through the next 50 years of biological research, but this is not possible. Perhaps they will allow me to return for the afternoon of April 26, 2053 (the 100th anniversary of the discovery of DNA), just to let me see what has happened. Will everything be solved by then? I don't think so, because problems have the curious property of reappearing in new guises as we come to ask new questions. I am also sure that if humanity survives, so will science, for while there are still humans with exploring minds, the quest for knowledge will continue.

Michael Brown

Medical research. Nobel Prize in Physiology or Medicine, 1985, for discoveries concerning the regulation of cholesterol metabolism.
Professor of Medical Genetics and Internal Medicine and Director,
Erik Jonsson Center for Research in Molecular Genetics and Human Disease, University of Texas Southwestern, Dallas.

My initial interest was in medicine, but when I was in medical school 40 years ago, it became clear to me that there were many unanswered questions in medicine and that the power of science to answer them was just beginning to be felt. I was infatuated by it.

I started as an amateur radio operator when I was 13, and that's what really kindled my interest in science. I used to build radio transmitters. It's similar to doing science in the sense that you make a circuit, try it out, and it doesn't work and you blow every fuse in the house and your parents get really upset. And then you have to go back to the drawing board and try to figure out why it didn't work and what connection you soldered wrong. Eventually, it works, and that's really the way science is!

I have a scientific collaborator, Joe Goldstein. He and I have worked together for 33 years. In 1969, we were both learning science at the National Institutes of Health outside of Washington. We saw two children. One was six and the other eight, a boy and a girl who were brother and sister. They were hospitalized because they were having heart attacks and couldn't walk across the room without having angina or chest pain, the kind of thing we normally see in 60-year-old men. The reason for their heart problems was that their cholesterol levels were over a thousand. At that time, there was nothing one could do for them. Coronary bypass surgery hadn't been invented. There was no such thing as heart transplants. They didn't respond to any medicines, and even if you put them on a zero cholesterol diet, their cholesterol remained elevated.

We decided we wanted to try to figure out what could possibly be wrong, what genetic problem could cause this high blood cholesterol. We both ended up in Dallas and worked together. We were fortunate enough to be able to figure out the problem. Unfortunately, it was too late for those two children, but other children subsequently have benefited from this work. On a broader scale, the work led to the basis of developing new drugs that lower cholesterol levels in people, and right now these drugs are being taken by more than 20 million Americans. It's very gratifying, since I was trained in medicine, to be able to do something that would help a lot of people.

Cholesterol itself is neither good nor bad, but it travels around the blood in packages complexed together with proteins. LDL is one of those packages, and HDL is another package. The problem with LDL is that it's very sticky, so when it circulates around, the package gets stuck on the walls of the arteries just like sludge on a boiler pipe. It gets taken into the artery and causes a big deposit. The dietary advice and all the things we've been doing to try to get people to lower their LDL levels has had an impact. The death rate from heart disease has gone down dramatically over the last 25 years. We are making progress. The new drugs that lower cholesterol are even better than diet in terms of lowering cholesterol. The studies have shown that even when you start these drugs late—that is, after people have already had a heart attack—there's about a 30–40 percent reduction in subsequent heart attacks. In fact, you can show that people actually live longer.

Richard Dawkins

Evolutionary science. Charles Simonyi Professor of the Public Understanding of Science, University of Oxford.

My early childhood in Africa should have turned me into a naturalist and biologist, but I don't think that was what did it. It was the philosophical end of science that more and more attracted me. I wanted to understand why we exist, why the world exists, why anything exists. At school, I drifted into the biology stream, probably in imitation of my father (a much better naturalist than me), and I found myself reading zoology at Oxford, where he had read botany.

It was not until my second year at Oxford that I took off, and I think it was its unique tutorial system that made me. I reveled in the privilege of a whole hour alone with a tutor every week. At least as important as the tutor is the regular discipline of having to write a long essay, using one of the world's great libraries: the thrill of following trails of original research literature. It now occurs to me that, having completed an essay on some narrowly defined subject, the Oxford undergraduate can temporarily fancy himself a world authority on it. Who else has read it up so recently or so thoroughly?

Later, as a tutor myself, I continued to enjoy this unique educational system from the other side of the fireplace. You find yourself teaching the same subject again and again to different individual students, and this hones your skills in the gentle art of explaining. As if by Darwinian selection, you feel yourself getting better at explaining each difficult idea. Which metaphors work? Which imagery and examples provoke the dawning smile of comprehension—which are rebuffed by blank puzzlement?

In my middle thirties, I tried to unite the explaining skills I was learning as a tutor with the writing skills I had practiced as a student, and I wrote my first book, *The Selfish Gene.* It is now seen as more novel and innovative than I thought it at the time. I believed I was just explaining orthodox Darwinism, but using unorthodox imagery which had proved its worth over ten years of tutoring. What I didn't know is that a new way of seeing science can be just as innovative as the science itself.

The Selfish Gene is not about selfishness. It presents a gene's-eye view of evolution. Individual organisms are temporary survival machines, throwaway vehicles for the immortal genes that travel through them from the remote ancestral past to the distant future (in the case of successful genes) or to oblivion (in the case of unsuccessful genes). Darwinian selection is the accumulation in the world of successful genes at the expense of their unsuccessful rivals. What we see when we look at individual organisms is the attributes that genes have built into them as tools to lever themselves into a long series of "next" generations. In a later book, *Unweaving the Rainbow,* I argued that an organism's genes can even be seen as a written description of the worlds in which its ancestors survived. *The Extended Phenotype* expounds other developments and extensions of the selfish-gene idea for a professional audience.

In *The Blind Watchmaker, River Out of Eden,* and *Climbing Mount Improbable,* I attempted to dissolve difficulties, not so much in understanding Darwinism as in believing that such a simple idea could do so much explanatory work. You can write it out in a phrase: the nonrandom survival of randomly varying coded information. That's it, the explanation for the whole of life, in all its magnificent diversity, beauty, and overpowering illusion of "design". Yet nobody thought of it until the19th century, and many cannot grasp it to this day. I have devoted much of my career to trying to dispel their difficulties. Honest difficulties, which I can respect, have been exacerbated by active and obscurantist opposition from religious lobbies, for which I am increasingly losing what respect I once had. My volume of collected essays, *A Devil's Chaplain,* brings together some of my polemics against religion, together with gentler pieces including travel writings, forewords, and memorials.

My other aim as a writer and teacher, implicit in all my books but explicit in *Unweaving the Rainbow,* has been to promote science itself as a cultural treasure on a par with music and poetry. We need musicians, and we need scientists. But you don't have to play an instrument to appreciate music, even at a high level. The equivalent ought to be true of science. Yet too many people are put off science at school because they associate it with Bunsen burners and pickled dogfish, the equivalent of five-finger exercises on the piano. The escape from the Bunsen burner should not be a dumbing down toward a travesty of science as larky "fun." Instead, science can become a romantic journey of enlightenment. We are privileged to live in a century where science can nearly answer those deep questions to which religion once vainly monopolized. What is the nature of the universe? Where did life come from, and what is its meaning? Now, before we die, we have the opportunity—and it is a privilege—to understand why we were ever born. It is only science that can even hope to answer the deep questions of existence, and I think it is a high duty of scientists to communicate the wonder of their subject, not just to students but to everybody who will (or even won't) listen.

Janet Conrad

Neutrino physics. Associate Professor of Physics, Columbia University.

You can't see them, but they're everywhere. There are 60 billion in front of your nose this second. You can't smell them. You can't taste them when they touch your tongue. You can't hear them. You can't feel them when they pass through you. Some 10,000,000,000,000,000 will do it while you read this page and you will never know.

They are neutrinos, the "little neutral ones" in the family of subatomic particles. They hold secrets from the earliest days of the universe. They bring us information from deep inside exploding stars and from high-energy particle collisions. Their presence may signal unexpected phenomena. Measuring their properties will help us understand how the universe will evolve.

We search for neutrinos using detectors all around the world. My experiment is at Fermi National Accelerator Laboratory, located just outside of Chicago. Our detector, MiniBooNE, is a 40-foot-high sphere, filled with mineral oil. When the neutrinos interact in the oil, they make small flashes of light that are seen by phototubes, which work essentially like inverse lightbulbs: light goes in, an electrical pulse comes out. The phototubes are 8-inch-diameter circles that are beautiful dark amber. When we installed them on the ceiling of the detector, MiniBooNE looked like a bizarre, beautiful planetarium with many moons and tiny stars that were the screws holding the tubes onto their black support structure. Despite being made of 800 tons of oil, MiniBooNE is actually quite mini compared with other neutrino experiments. The most awesomely large neutrino detector, Super-K, is located in Japan. A 15-story building can fit comfortably within that detector!

We need such large detectors because neutrinos don't interact with matter very often. Most subatomic particles are very interactive. For example, quarks, which make up most of ordinary matter, are so active in our detectors that it is difficult to sort out the patterns that they leave. The electron is another highly evident particle—and reliable, too. You can count on finding electrons inside your typical wall outlet and also inside your typical particle interaction. But the neutrino is different from the rest. Their interactions occur far more rarely. At the highest-energy accelerator in the world, Fermilab, we observe neutrino reactions 10 billion times less often than those of quarks. They just quietly zip through the detector and go on their merry way.

Neutrino research is fascinating because today's results are full of contradictions. For 50 years, all of the evidence pointed to neutrinos being bundles of moving energy that had no mass—a pretty weird concept for a particle. But recently we discovered a novel behavior which can be explained only if neutrinos do have mass. How to resolve this conflict?

If the neutrino has mass, it must be very, very small. It would take at least half a million neutrinos to tip the scales on the electron. Still, such a wispy particle will have a big effect in the universe. The collective mass of the neutrinos rivals the mass of all the stars! Given the discovery of mass, we can begin asking even more exciting questions. The Big Bang, for example, produced a million neutrinos in every gallon of space. The holy grail of neutrino physics is to detect these relics. Their mass may hold the key.

All of that sounds pretty esoteric, and you may ask, "What have neutrinos done for me lately?" Actually, they matter a lot to you. They are part of the ignition process of the Sun. They play a role in heating the center of Earth, causing continental drift. So the next time you see a koala, whose evolution depended on living on an isolated continent, thank a neutrino! The tools that physicists use to create and study neutrinos have direct benefit to every one of us. One fork of the beam line for our neutrino experiment at the Fermilab goes to neutron therapy, a very successful cancer treatment method. The extremely clean laboratory environment of state-of-the-art solar neutrino experiments can be used for sensitive tests to monitor violations of the Nuclear Test Ban treaty.

I love this particle. It always reminds me that even the smallest among us can change the universe.

James Darnell

Molecular cell biology. Vincent Astor Professor, Emeritus, Rockefeller University.

Reflections on the route by which an individual became a scientist necessarily produce insight into the culture into which the person was born. In a small Mississippi town (Columbus) in the 1930s there were no scientists, nor, as I recall, awareness of science. Even my high school teachers, though earnest, were largely unschooled. And no one in my family was in any way acquainted with science. My mother was a dietitian in one of the two local hospitals (each incidentally *owned* by a prominent local practitioner), and by taking lunches at the hospital I had contact with the five or six physicians who used the hospital. I saw at first hand that young southerners of modest background could in fact get their education in far-flung and prestigious places and, not an unimportant consequence, get out of the "old South." Through acquaintance with these physicians, I chose to head into medicine, and after a B.S. at the University of Mississippi, I enrolled, for the first year on a hometown loan, at Washington University (St. Louis) School of Medicine in the fall of 1951. Subsequent years were supported by scholarship.

I was introduced to laboratory science in my second year of medical school when each student was, in lieu of the standard microbiology laboratory, assigned to a professor's research team. My extraordinary good luck with mentors began here. Robert Glaser, who studied streptococcal disease, allowed me free range in his laboratory, and through some experiments concerning streptococcal susceptibility to penicillin only during rapid growth, I gained an entree into Harry Eagle's laboratory at NIH for post-M.D. training. Eagle had earlier studied penicillin, but by the time I joined him he was studying conditions for successful growth of animal cells—human and mouse—in culture.

In Eagle's group, a successful set of studies on poliovirus infection of human cells earned me acquaintance with the great virologist Salvador Luria, a one-year postdoctoral period in François Jacob's laboratory at the Pasteur Institute, and a beginning job as an assistant professor at MIT in Luria's subdivision in the Department of Biology. While I would not deny that captivation by science, devotion to science, and hard work were instrumental in my arriving at this point, I've never doubted that the most marvelous good luck had carried me to the contacts with Glaser, Eagle, Jacob, and Luria and got me started in June 1961 in my own laboratory at MIT.

At this time, experiments with bacteria had unlocked many of the major secrets of how the gene in bacterial cells instructed the synthetic machinery to make all the required proteins of the cell. Did these principles apply to animal cells? We were emboldened to take a crack at this question, and my scientific life unfolded along this path. Along the way—at MIT, then at several New York institutions—we found that getting the instructions out of DNA into RNA in animal cells presented problems not encountered in bacteria. And it is, after all, absolutely necessary for cells to recover information from the essentially inert DNA by transferring the information to RNA, the real director of the cellular ballet. The primary RNA copies of DNA in animal cells had to be modified through a number of surprising steps of molecular carpentry, the most amazing of which was the cutting into pieces of an initially very long string of RNA, discarding many pieces, and then rejoining of selected pieces. This feat is accomplished in human cells in the production of virtually every messenger RNA molecule—the type of RNA that carries instructions for making a protein.

Biologists with a medical background toiling in the swamps of ignorance in the 1950s and 1960s never realistically expected their labors to be useful to medicine. By the 1990s, the climate had changed completely. Expectations, also sometimes unrealistic, arose that much of molecular biology would prove useful in medicine. Our own work in the 1990s and continuing to the present deals with protein actions that are used in most cells of the body to respond to signals delivered from outside the cell. Faulty regulation in these signaling pathways leads to many diseases, including cancer. Information about drugs that directly affect such unregulated proteins can be and is being sought, with the hope of benefiting patients.

If a scientist, at least a biological scientist, is gifted with good health, life continues to hold challenges and unfolding mysteries that tantalize and amuse. We biologists are not necessarily ushered to the sidelines by age. Can there be a better course to follow in life?

Rita Colwell

Biotechnology. Chairman, Canon U.S. Life Sciences. Former (11th) Director, National Science Foundation (1998–2004). Professor, University of Maryland and Johns Hopkins University.

I grew up in a small town in Massachusetts, Beverly Cove, a block from the lighthouse overlooking Beverly Harbor. I attended the public schools in Beverly Cove and Beverly town. They were both traditional and very good, particularly because of the teachers. There were two significant turning points in my life. One was the day the principal, Miss Striley, a kind but very stern woman, called me into her office. A student got called into Miss Striley's office only for punishment. This day, however, Miss Striley said, "The results of the national tests you and your classmates took recently have just arrived. Your grades were the highest ever recorded in this school." Wagging her finger at me, she added, "You must go to college!" The second pivotal experience was getting to know my brother-in-law. He and his friends, all of whom became internationally recognized physicists, made informal visits to our home. This fascinated me so much that I chose to study science when I went off to college myself. Ultimately, I became a molecular microbiologist.

Marine species, both large, such as fish, and small, like zooplankton, carry bacteria of the *Vibrio* species. When I was an assistant professor at Georgetown University, a friend at the National Institutes of Health suggested I study the cholera *Vibrio* because of my extensive experiences with marine *Vibrios*. *Vibrio cholerae* had an absolute requirement for salt and broke down chitin, the whitish, hard substance that crab shells are made of. It then became clear to me that *Vibrio cholerae*, the cause of devastating epidemics of cholera in humans, was a bacterium that lived naturally in rivers, estuaries, and coastal waters. This was totally against the dogma of the times (the 1970s). My students and I showed that this same bacterium was an inhabitant of Chesapeake Bay, along with other *Vibrio* species, contrary to the dogma that *Vibrio cholerae* was transmitted only from person to person and had no life in the environment. We did very careful DNA analyses, using the most advanced methods of the time and showed that, indeed, our isolates and clinical isolates from cholera victims were not only closely related but the very same species!

Another finding was that *Vibrio cholerae* possesses a dormant stage in its life cycle. Using molecular genetic tools, we were able to demonstrate that the bacterium goes into a stage where it cannot be cultured in the laboratory, but remains viable. Since that discovery more than 20 years ago, there have been more than a thousand papers showing that many bacteria demonstrate this same phenomenon—namely, that under conditions not conducive to active growth and reproduction, bacterial cells will "hunker down" until conditions improve. Most importantly, we discovered that *Vibrios*, notably *Vibrio cholerae*, are associated with zooplankton. These microscopic marine animals, the zooplankton, are abundant in the spring and the fall.

Vibrio species are common to pond, river, estuary, and coastal waters. Cholera epidemics in Bangladesh show remarkable seasonality, year after year. Using satellite remote sensing to measure seawater temperature in the Bay of Bengal, we found that there is a peak in sea surface temperature in the spring and the fall of each year. The phytoplankton (microscopic plants in the sea) increase first, responding to the warmer seawater temperature and increased sunlight, and then the zooplankton (microscopic animals) burst into a "bloom," feeding on phytoplankton. We showed that cholera epidemics and the sea surface temperature increases were linked.

I concluded that if we could remove the plankton from the pond and river water that the villagers in Bangladesh drink, we could reduce the incidence of cholera. In the laboratory, we did experiments showing that filtering water removed copepods, and *Vibrio* counts dropped by 99 percent, because the *Vibrios* were attached to the copepods. We then carried out a three-year study in Bangladesh that included about 65 villages, comprising 144,000 people, with villages divided into those where inhabitants filtered and those where they didn't. The filter we devised was a yard of old sari cloth folded about four to six times, ideal for villagers who had very little money, but did have access to used sari cloth. When we analyzed the results, cholera was reduced 48 percent in villages that filtered their water.

I have always been interested in environmental factors associated with human health and began my work in Bangladesh in 1976. All but a couple of the directors over the 25 years I worked at the International Center of Diarrheal Diseases Research believed I was pursuing an implausible hypothesis. Some even made my life pretty miserable, but I just continued to do the work, gathering sound data, analyzing the data carefully, and publishing prolifically. I never gave up and always found a way to patch together resources and get the work done. When I come up against unfair opposition, it just makes me work harder to prove my opponents wrong and focus on a strategy to overcome the impediments.

Robert Edwards

Reproductive biology. Professor of Human Reproduction, Emeritus, Churchill College, University of Cambridge.

Shortly after my army demobilization in 1951, a university training in genetics beckoned. Desiring to study developmental genetics in mammals, I did my doctoral thesis on chromosomes in mouse embryos. After four years, numerous mouse embryos with haploid, triploid, tetraploid, and other chromosomal disorders had emerged, and I had mastered reproductive biology and genetics, then a very rare combination of subjects. Geneticists and reproductionists did not talk together or share conferences. Yet the immense dual potential of my research for human studies and its ethical implications was already stimulating me.

The clinical significance of reproduction and genetics was reinforced in the late 1950s as some patients with Down syndrome and other disorders were found to inherit nondiploid chromosomes. Human disease genes were also being clarified. My background had trained me to study these disorders in mouse embryos, and in human embryos too, provided that human fertilization was achieved in vitro, since it might be possible to identify and discard early carrier embryos. I had also become fascinated by growing embryo stem cells in vitro, and collaboration with John Paul and Robin Cole on rabbit embryos in Glasgow led to the production in vitro of immortal stem cell lines in vitro, and of cells differentiating into blood, muscle, nerves, phagocytes, connective tissue, and all the body's fundamental organs.

Discovering some friends who were infertile made me realize I might help them by fertilizing human eggs in vitro to prepare embryos for transfer to their mothers' uteri. A fantastic program, attracting hostile ethical attention, promised to avoid genetic disorders in babies, produce human embryos to cure infertility affecting almost 10 percent of married couples, and grow stem cells to mend disordered body organs. What a program for a young scientist with no medical background and facing intense ethical opposition! I never accepted any religious objections, convinced that the earliest stages of human life must come under the care of science and medicine. But only if human embryos could be developed in vitro!

My first medical colleague, Molly Rose, of the Edgware Hospital, London, sent me slithers of human ovaries. Their few immature human oocytes could be aspirated for study. After three years' travail, I mastered their maturation in vitro; i.e., they were ready for fertilization. Some years later, human fertilization in vitro was achieved, in 1969, in work with one of my graduate students, Barry Bavister. With Richard Gardner, another graduate student, a few cells were excised from rabbit embryos and used to correctly diagnose the sex of full-term embryos. Pre-implantation genetic diagnosis had begun and has spread today to designer babies. Ethics and dealing with the press were now major aspects of my life. For his Ph.D., I asked Richard to take a single mouse stem cell and test its properties by injecting it into a recipient blastocyst. It contributed descendant cells to virtually every tissue in the body, matching those we had grown in vitro. Stem cells now promised immense curative properties, a matter of intense interest to me even though no one else in the world was interested. I met my long-term medical partner, Patrick Steptoe, and the drive to produce "test-tube" babies began a ten-year travail to the birth of Louise Brown, the world's first. The immense ethical, scientific, and clinical problems of those years will never leave me.

I had reached the end of the beginning of my research. Human embryos were available for genetic diagnosis or to make stem cells. In vitro fertilization promised to alleviate a great deal of infertility. Hundreds of babies proved my contention that IVF-associated risks would be minimal, and embryos used to make stem cells were normal.

Christian de Duve

Cell biology. Nobel Prize in Physiology or Medicine, 1974, for his discoveries concerning the structural and functional organization of the cell. Andrew W. Mellon Professor Emeritus, Rockefeller University. Emeritus Professor, Catholic University of Louvain, Belgium. Founder-Administrator, Christian de Duve Institute of Cellular Pathology, Brussels.

Born in England during the First World War, of Belgian parents with partly German roots, I grew up in the cosmopolitan city of Antwerp, where I had the benefit of a classical education taught in the two national languages of Belgium, French and Dutch. By the time I entered the Catholic University of Louvain, in 1934, I had become familiar with two more languages, thanks to stays with English friends of my family and with German relatives; and I grandiosely called myself a "citizen of the world," labeling as hopelessly passé my father's growing worry over the revival of German nationalism. Unfortunately, as fathers often are, he proved to be right.

I watched the rising tragedy from a distance, having meanwhile discovered what was to be the passion of my life. Although attracted by the humanities, I had chosen medicine as a career, seduced by the image of the "man in white" dispensing care and solace to the suffering. But science was lurking around the corner, in the form of an unpaid student assistantship in the laboratory of physiology. There I was allowed to do a little work on insulin. I promptly fell in love with scientifc research and soon had assigned myself, as a major vocation, the task of elucidating the mechanism of action of the antidiabetic hormone.

The sound of Stukas dive-bombing the university library in the early hours of 10 May 1940 shattered my complacency, but not my dream. After a spell in the army and a not very glorious escape from a prisoners' column, I went back to work in occupied Belgium, with only a mild participation in the local resistance movement, combining a clinical internship in a cancer ward with the study of chemistry and the writing of a monograph covering all the research that had been done on insulin since its discovery in 1922. Somehow, all this was achieved by the time the war ended. I was able to complete my training in biochemistry in two prestigious laboratories, one in Sweden and the other in the United States, where I had the privilege of working with no fewer than four future Nobel laureates. I learned from them how to do the kind of technically impeccable and intellectually rigorous research that occasionally, with a great deal of luck and a nudge from serendipity, earns a Nobel Prize.

Serendipity did call, in the late 1940s, in the small laboratory I had started at Louvain University, presenting me with a strange observation that tickled my curiosity and led me to abandon, for what I thought would be only a brief interlude, my chosen goal to solve the insulin conundrum. At the end of the trail, fifteen years later, two cell organelles had been characterized. I called them lysosomes (digestive bodies) and peroxisomes (metabolizing hydrogen peroxide), for reasons related to their biochemical functions. It took another ten years for these lucky findings to be singled out for the 1974 Nobel Prize in Physiology or Medicine, which I shared with my countryman Albert Claude and his Romanian-American former collaborator George Palade, two pioneers in cell biology.

In the meantime, I had in 1962 been offered a professorship at the celebrated Rockefeller Institute (now University) in New York (where both Claude and Palade did their most important work). I was able to accept this offer without abandoning my Belgian laboratory and have since then shared my time between the two institutions. In addition, with a few colleagues, I created in Brussels a new biomedical research institute, now known, against my protests, as the Christian de Duve Institute of Cellular Pathology, or ICP, where, since its opening in 1974, some 200 investigators try to follow the Rockefeller ideal of joining first-class basic research with the development of beneficial applications "pro bono humani generis."

All these activities are now behind me. They have yielded to a new, burning interest in the origin and evolution of life and in the philosophical and religious implications of our newly acquired knowledge in these all-important fields, the subject of my two most recent books, *Vital Dust* (1995) and *Life Evolving* (2002).

I have had the good fortune to live, as an inside witness and, even, a modest participant, at a time when our understanding of this wonder we call "life" has made its most revolutionary advances. Today, in the dusk of a long life, I look back with gratitude toward all those—my wife, Janine, my family, mentors, co-workers, colleagues, friends, and, not to be ignored, Lady Luck—who have helped me in this exciting and fulfilling adventure. I contemplate with a mixture of anxiety and confidence the ways in which future generations will use the knowledge and power ours has gained for them. May they do it with more wisdom than is common in our own day and age.

Mary Eubanks

Plant biology, evolution of maize. Adjunct Professor of Biology, Duke University.

Who would have ever imagined that my anthropology major would one day lead to a career in plant genetics! But I was well prepared because, growing up in Mississippi, where my family owned the world's largest pecan nursery, I got hands-on experience in plant breeding at an early age from my grandfather, who developed several new varieties still popular today. My roundabout path into science was through graduate research in archaeology that focused on maize and pre-Columbian pottery. The maize ears depicted on ceremonial jars were not primitive artistic renderings. They were molded from impressions of real ears, and as such are fossils preserved in clay that permit identification of indigenous races and provide a unique window into evolutionary history, illustrating the rich diversity created by ancient growers of maize that was traded great distances thousands of years ago.

Although many experts believe the wild grass teosinte is the sole progenitor of maize, that hypothesis does not accommodate essential evidence linked to its origin. An unusual feature on Zapotec urns from the region of southern Mexico believed to be the cradle of maize originally had me quite perplexed. Attached to the tips of the mold-made naturalistic ears is a stylized element that resembles a staminate spike, the plant part bearing the pollen-shedding male flowers. This was puzzling because the male and the female flowers are borne on separate parts of the plant in corn and teosinte—male flowers in tassels at the ends of the stalks and female flowers in ears on the sides of the plant. However, corn's other wild relative *Tripsacum*, common name gamagrass, has the male and female flowers together on the same spike. This feature is also seen in the oldest archaeological specimens from Tehuacán, which adjoins Oaxaca.

The primary obstacle to elucidating the question of the origin of maize has been that no one could experimentally synthesize the transformation of the tiny, shattering teosinte spike with a few seeds into corn's phenomenal reproductive structure, which is unparalleled in the botanical kingdom, the many-rowed ear with hundreds of kernels. Then, in the late 1970s, a perennial teosinte was discovered on the verge of extinction that had the same chromosome number as corn, but the primitive chromosome architecture of *Tripsacum*. As a postdoctoral fellow working under the world-renowned maize cytogeneticist Dr. Marcus M. Rhoades, I recognized the unique opportunity this new teosinte provided for experimentally crossing teosinte with gamagrass, and the hybrids I made provide an important breakthrough for basic and applied research. Prototypes of archaeological specimens recovered after a few generations of experimental crosses demonstrate how corn could have evolved rapidly from human selection and cultivation of natural intergenomic recombinants between gamagrass and teosinte. Since the gamagrass-teosinte hybrids are cross fertile with corn, they provide a genetic bridge to move hardy genes from gamagrass into corn using conventional plant-breeding methods. Skepticism and dogma are hallmarks of science that slow advances from seminal discoveries. Though still ignored by certain experts, a paradigm shift in the way many scholars are thinking about the origin of maize has begun. Though the seed industry's dominant technology is currently genetic engineering, the ability to produce corn that is naturally resistant to the billion-dollar bugs (corn rootworm and European corn borer), and has the strong drought tolerance of gamagrass, offers a cost-effective and noncontroversial alternative to genetically modified corn.

Walter Gilbert

Biochemistry, molecular biology. Nobel Prize in Chemistry, 1980, for contributions concerning the determination of base sequences in nucleic acids. Carl M. Loeb University Research Professor, Harvard University.

As a child, I was interested in chemistry, in hunting for minerals, and in astronomy. I always thought I would be a scientist. I went through high school playing hooky to read up on science in the Library of Congress. In college, I was a chemistry and physics major, but I then became a theoretical physicist and was an assistant professor of physics at Harvard when, in the spring of 1960, I began doing molecular biology.

I followed Jim Watson (we had become friends in 1955) around one day as he was doing an experiment, and the next day I joined in. That experiment was to detect an unstable intermediate, messenger RNA, that's made as a copy of a gene along DNA and that carries information out to the factories in the cell that make proteins. After making a few copies of the protein, each messenger is destroyed and a new one used.

I became a molecular biologist to ask the question "How are genes controlled?" In bacteria, the way in which genes turn on and off is that one gene makes a protein and that protein binds to the DNA of a second gene and turns it off. Benno Mueller-Hill and I were the first to isolate such a protein, a repressor. In higher cells, DNA binding proteins turn genes both off and on.

What were the DNA sequences—the order of chemical groups, the bases, along the DNA—to which this protein bound? By using the very laborious techniques that were available in the early 1970s, Alan Maxam and I worked out a sequence of 20 bases for the target of the repressor, 20 units along the DNA, and it took us approximately two years, about a month per base, to work out that little structure. If there are 1,000 bases in a gene (as we thought then; we now would say 10,000 bases in a gene for other reasons), it would take one a century.

In the middle 1970s, we discovered a very rapid way of working out the DNA sequence—so rapid that in an afternoon a single person could read a hundred to a thousand bases of DNA. The method breaks up the DNA into pieces whose lengths determine the positions of the different bases. When these pieces are separated by size, their positions enable one to read the sequence. This method, and a similar one discovered by Fred Sanger, transformed biology. The rate of DNA sequencing now increases tenfold every five years, a millionfold since 1975.

In 1974, one couldn't know the sequence of a gene. By 1975, that was easy. By 1985, I could say, "We are sequencing so much DNA that we could imagine knowing all human genes." That suggested the Human Genome Project, and by 2000 we had a good rough draft of the entire human genome. The DNA sequence of the entire genome is essentially a stake in the ground that says, "Well this is all there is. These are all the human genes." One cannot yet interpret all the relevant information from the DNA, but now one can decipher what each piece is and what it's doing. Biology has changed from a one-gene-at-a-time approach to "which of the 40,000 possible human genes are involved."

Val Fitch

High-energy particle physics. Nobel Prize in Physics, 1980, for the discovery of violations of fundamental symmetry principles in the decay of neutral K-mesons. James S. McDonnell Distinguished University Professor of Physics, Princeton University.

In my next incarnation, I want to be a great black-backed gull. From the deck of our summer home on the shore of Atlantic Canada, I have watched these majestic birds for many hours soaring on the wind currents sometimes to great heights. They are not interested in going from A to B, nor are they looking for food. They are soaring for the pure joy of it—clearly a wonderful, exhilarating experience for them. But I have been asked to say something about my present incarnation.

In retrospect, the trajectory of anyone's life is never a straight line. However, it does appear especially improbable to have been born on a cattle ranch in northwestern Nebraska and 57 years later appear in Stockholm to receive a Nobel Prize. My father was a cattle rancher who raised purebred Herefords, and my mother had been a local schoolteacher. I had a brother ten years older and a sister six years older. On the ranch, they got on their ponies and went up the valley to a one-room schoolhouse. However, when I turned five we moved to the nearby town of Gordon, Nebraska, and it is there that I went to school, K–12.

I cannot say that I had any role models. Science was always in my blood. My father fashioned a bench in a corner of the basement, and that is where I did my "experiments." My family was extraordinarily tolerant of the fumes, pops, and sparks that came out of my laboratory. I was finally requested to install my own fuse box to save the rest of the house from darkness when my electrical experiments got out of hand. Among my experimental gear, I had the usual chemistry set, an Atwater Kent radio, ca. 1930, a spark coil from a Model T Ford. On the ranch my mother, in the absence of electricity, had a washing machine powered by a gasoline engine. I used that engine to build go-carts and lifting devices. It was a rich life for one interested in how everything worked, in what made the world tick. Around the age of 12, I learned there was a profession devoted to such things. It was called physics.

It was my older brother who really led me out of the wilderness. When I was graduating from high school, he was working on his Ph.D. in economics at Columbia University. Letters from him broadened my interests in literature, in world affairs, and in politics. It was he who first pointed me to the delightful essays of E. B. White in a column entitled "One Man's Meat" appearing in *Harper's Magazine*.

During all these years, my father continued ownership of the ranch and he would have loved to have had one of his sons take over. After World War II (during which I had spent three years in the army), it became clear to him that neither son was remotely interested in the ranch and he sold it.

The Nobel Prize was awarded for the discovery of a reaction that violates an important symmetry principle, that of time reversal invariance. While it appears to be a highly abstract concept, it has a profound effect. The universe began (the Big Bang) with the production of equal numbers of particles and antiparticles. Most of these were subsequently destroyed through their mutual annihilation. The result of the annihilation is the cosmic background radiation that fills our universe. However, because of a lack of time reversal symmetry, a slight excess of particles was left over, roughly one in ten billion. This is the stuff out of which all is made. If this small asymmetry had not existed, we, and the universe as we know it, would not exist.

William Schopf; Jane Shen-Miller Schopf

Evolutionary biology. Professor of Paleobiology, University of California, Los Angeles; Ancient living seeds. Research biologist, University of California, Los Angeles.

Though recollections of my youth have been blurred by the passage of time, I remember well the first day of my fourth-grade class in Columbus, Ohio. Our teacher (Ms. Tinapple, as I recall) asked each of us to report to the class what we "wanted to be" when we "got big." The girls' goals—to be teachers, secretaries, or nurses—all focused on helping others. Some of the boys wanted to be basketball or football stars. When it came my turn, I announced without the slightest hesitation, "I want to be a professor." (As I think back, it strikes me as interesting that I chose "professor" rather than "scientist"—probably because it seemed impossibly bold for me to follow in my father's footsteps and become a "real scientist.")

My dad was a professor (paleobotany) at Ohio State, and my mom was schooled in botany, mathematics, and several languages. From an early age, it was a "given" that both I and my older brother would earn doctorates. Whether a result of nature, nurture, or both, I did end up a professor and a scientist both. Spurred on by geology professor Larry DeMott as a sophomore in college, I became fascinated by the question that has formed my life's work: "When and how did life on Earth get started?" In those days, the oldest fossils known (lobster-like trilobites) dated from about half a billion years ago. But as Darwin had pointed out when he unveiled his theory of evolution, such trilobites were far too complicated to have anything to do with life's beginnings. He was deeply perplexed by the missing record of life's earlier history, and thought this serious absence could "be truly urged as a valid argument" against his theory. By the time I entered the scene, the search for the missing early fossil record had become notorious as a burial ground of wrongheaded discoveries. Yet I was young and enthusiastic, so I set myself the goal of tracing life's roots back through time in hopes of finding life's beginnings.

Progress has been made. The oldest records of life now date from about three and a half billion years ago, a sevenfold increase from what I was taught in college, and I've had a hand in setting up a brand-new field of science. A great deal has been learned about how the earliest microorganisms evolved over time to the flora and fauna of today.

Until the1980s, anti-evolutionists routinely bolstered their claims by quoting Darwin's worries about the "missing early fossil record." Though they never recanted, they have stopped using this line of argument. The record of early life, and the evidence of evolution it presents, had become irrefutable. Darwin would have been pleased.

Born in China, one of six children of an educated family, I was taught that to be a teacher was the highest position a person could ever achieve. One winter night, my parents told us a story about Dad's cousin, a Ph.D. wheat geneticist. "Children, the dumplings you had for dinner came from wheat. Until a few years ago, our wheat stalks had been easily damaged by storms from the Gobi. To improve the wheat, your aunt and uncle painted wheat flowers one-by-one with fresh pollen of hardy stock. One morning, after a night storm and rain, they saw among a massive field of the weakling plants a sturdy bunch of baby 'wheatlings' standing straight and strong. Since then, their new wheat has been spread all over the Northwest, and wheat production has doubled, tripled, even quadrupled!" We sisters, being particularly fond of dumplings, clapped with glee.

I studied agriculture in college, at a time when it was not a popular subject for a girl, earned a doctorate at Michigan State, and have had a satisfying career in plant science. Some years ago, I was fortunate to become interested in the Sacred Lotus (variety *China antique*), the showy water plant that in Buddhism symbolizes purity, emerging from the mud to rise high above the water. Used by Chinese doctors for over 4,000 years, lotus seeds and stems are also delicious, fresh or cooked.

I began my work with seven seeds from an ancient lake, now dried, in Liaoning Province, northeastern China. After storing them for 10 years in my lab, I tested their germination. The oldest that sprouted dated from 1,300 years ago (as shown by ^{14}C radiocarbon). Remarkably, this still is the oldest living seed in the world. To sprout, it had to repair hundreds of years of stress. Understanding this repair could provide insight into the aging process in *all* biology, including our own.

How does a lotus seed maintain its unsurpassed longevity? Its greatest asset is an outercoat, impervious to air and water. Lotus seeds can be soaked in water for a year or more, sprouting only if their coats are cracked. The coats also contain chemicals that prevent bacterial and fungal invasion. Unlike known proteins in embryos of other plants, 60 percent of those in lotus embryos are heat hardy, even above the boiling point of water. These proteins, and other biochemicals, seem to play important roles in repairing age-caused damage.

In 2002 a Paris TV documentary invited my husband and me to film an hourlong episode on lotus, *Des Graines d'Éternité*. In far northeastern China, we visited a lake planted with *China antique*, offspring of the ancient seeds, where alluring lotus blooms again adorn the landscape as they did a thousand years ago.

Paul Hoffman

Earth surface history. Sturgis Hooper Professor of Geology, Harvard University.

I was lucky to discover geology at an early age, and it has never let me down. Like many youngsters, I was introduced to fossil hunting and mineral collecting by the staff of the local public museum. Taking myself more seriously than most, I decided I should be a geologist for life. My parents believed that children ought to be kept outdoors as much as possible and encouraged my resolve. Field geology, science on the hoof, linked the worlds of Robert Louis Stevenson and George Gamow[1] in my preadolescent imagination.

At the end of my freshman year in college, I went to the Ontario Geological Survey seeking a summer job and was told to buy a good pair of boots, take the train to Sioux Lookout, and prepare for four and a half months of canoe reconnaissance mapping north of Lake Superior. That summer, the deal was sealed. But later, as I sat taking classes, the great problems of crustal geology, the targets of my ambition, were solved in one fell swoop by the theory of plate tectonics. What was left to do? Plate tectonics had clearly operated ever since abundant fossils first appeared, but what about the prior 88% of Earth history? I signed on with the Geological Survey of Canada to find out. I was privileged to map, largely on foot, a 50,000 km^2 swath of uninhabited (and uninhabitable) terrain in the northern Canadian Shield, the foundations of the oldest continent on Earth. Many aspects of the Earth's surface can be effectively mapped from space, but interpretive continental geology is not one of them. It requires that eyeballs be carried inches from the landscape, with a rock hammer at the ready. In the modern world, field geology is as close as one gets to being a "hunter gatherer." As a species, we evolved through natural selection as hunter gatherers. It follows that we are biologically optimized to be field geologists.

What I found was that plate tectonics has operated for over half and possibly most of Earth history. The earliest record is too fragmentary to give a clear picture, so I turned my attention forward in time. Deposits of glacial origin are found in strata not far below the first abundant fossils on every continent, including continents then astride the equator. A colleague in California hypothesized that shelves of glacial ice had covered the oceans from pole to pole, creating what he termed a "Snowball Earth." By reflecting Sunlight back to space, the uncontrolled growth of ice itself was responsible for the climatic disaster. Isolated from atmospheric oxygen and winds, the ocean water beneath the ice stagnated, and marine life hung on by a thread. Ultimately, plate tectonics saved the day. Carbon dioxide emitted by volcanoes could only accumulate in the atmosphere of the frozen planet, eventually melting the ice by means of the greenhouse effect. My California friend published his Snowball Earth hypothesis as a two-page paper within in a 1348-page book. It was not an effective strategy to convince a skeptical audience of a radical idea.

Intrigued but deeply skeptical, I mounted a field project in newly-independent Namibia, in southwest Africa, where the glacial deposits in question are extensively exposed. It took six years of arduous field work and thousands of isotopic measurements of rock samples in the laboratory to convince myself that the predictions of the Snowball Earth hypothesis were true, and that no less radical hypothesis would suffice. Wrestling with the hypothesis had been the most intense learning experience of my life, aided immeasurably by my Harvard colleague, the geochemist Daniel P. Schrag. But a more difficult task lay ahead, winning over the majority of geologists who had ignored or scorned the hypothesis. My skills are in geology, not politics, and the battle is still ongoing. The hypothesis has been repeatedly challenged, and in the process it matured and became stronger. Yet, success is by no means assured, and like any scientific hypothesis, it could be supplanted by a superior idea at any time. What is certain is that there is no better life than one devoted to learning and extolling our planet's own story, nor one that brings greater reverence for our home in the universe.

1. Author of numerous popular books on science, George Gamow was a nuclear physicist, cosmologist, and biochemist who named and championed the "Big Bang" theory of the origin of the universe.

Murray Gell-Mann

Particle physics. Nobel Prize in Physics, 1969, for his contributions and discoveries concerning the classification of elementary particles and their interactions. Distinguished Fellow, Santa Fe Institute.
Robert Andrews Millikan Professor of Theoretical Physics, Emeritus, California Institute of Technology.

I grew up in Manhattan. A great deal of what I learned, I learned from my older brother. We both loved nature, and we enjoyed bird-watching, botanizing, and catching butterflies. My interests extended beyond natural history to human history, as well as linguistics, archaeology, and other parts of anthropology. All those subjects have to do with diversity, complexity, and evolution. They aren't much like elementary particle physics, but that is the field in which I worked for 40 years or so, before returning to my old interests.

It used to be thought that the neutron and proton—the constituents of atomic nuclei—were elementary (that is, fundamental) objects. I suggested in 1963 that they were in fact composite, made up (roughly speaking) of three quarks each. Quarks were hitherto unsuspected particles, with fractional electric charges of plus two-thirds and minus one-third in the usual units where the proton's charge is plus one. I also suggested that the quarks were stuck permanently inside directly observable objects such as the neutron and proton.

That the neutron and proton were not elementary was a radical idea. The fractional charges and the confinement of quarks were also radical. The difficulty in putting forward this scheme lay not so much in coming up with it, since it was suggested by patterns in the data on objects discovered experimentally in the fifties and sixties, but in believing it. The quark idea contradicted so many accepted ideas.

I contributed, along with some other physicists, to the construction of a quantum field theory of the quarks and of the "gluons" that hold them together. The theory is called quantum chromodynamics because the quarks come in three "colors," as I called them. (That property, which is associated with the gluonic force between quarks, has nothing to do with ordinary color—the name is really a joke.) Gluons, by the way, are confined inside directly observable objects, just as quarks are.

Pure science often has practical applications typically arising decades after the theories are stated and verified. But pure science also has a way of affecting people's lives through the way one thinks about the world. The theory of biological evolution has profoundly affected our thinking, of course, but so has the discovery that the universe has been expanding for around 13 billion years, starting as a tiny hot ball. Special and general relativity and quantum mechanics have changed dramatically the way humans understand physical reality. When the unified theory of all the particles and forces is finally in hand, it too will change our vision.

Sheldon Glashow

Theoretical physics. Nobel Prize in Physics, 1979, for contributions to the theory of the unified weak and electromagnetic interaction between elementary particles, including the prediction of the weak neutral current.
Arthur G. B. Metcalf Professor of the Sciences, Boston University.

I was the last child of immigrants from czarist Russia, born when my two brothers were teenagers. One would become a dentist, the other a physician, both serving as officers in the U.S. Army during World War II. As a myopic, overweight, athletically disadvantaged, but much spoiled child, I became an avid reader of action comics and fantasy. News of the atomic bomb turned me from a science fiction fan to an aspiring scientist. "Why not become a doctor and do science in your spare time?"' my plumber father would sometimes ask, but he was not too chagrined when I declined. Tiring of trapping exotic fish at our beach home and breeding hamsters in my bedroom, I was given a toy microscope, with which I found intestinal worms in the Hudson River and identified various protozoa from debris in the park across from our Manhattan home. When my interests turned to chemistry, my father built me a primitive basement laboratory. There, with parents upstairs and mercifully unaware, I created multicolored explosions and synthesized many noxious compounds of my favorite element, selenium. When my high school curriculum moved on to physics, so did I. At the Bronx High School of Science I found others with interests in science and mathematics. In subways, in bookstores, and on the telephone, we taught ourselves the rudiments of modern physics.

As a Cornell freshman, I was committed to theoretical physics as a career. At Harvard I became a graduate student of Julian Schwinger, who posed the problem that would lead to my Nobel Prize: Could the weak and electromagnetic forces have a common origin? In my 1958 thesis, I argued that a sensible theory of the weak force had to describe electromagnetism as well. In 1961, as a postdoc in Copenhagen, I identified the correct algebraic structure of such a model, but I could not explain what made the weak force so much weaker than the electric force. This obstacle was overcome by my high school buddy Steve Weinberg in 1967 with his ingenious use of spontaneous symmetry breaking. The last piece of the puzzle (the experimental absence of certain forms of radioactivity) was resolved by my introduction of the charmed quark. This then hypothetical particle enabled the so-called GIM mechanism, which I developed with John Iliopoulos and Luciano Maiani.

By 1974, with the twin discoveries of neutral currents and of particles made of charmed quarks, the electroweak synthesis became an integral part of today's standard model of elementary particle physics. At about that time, my colleague Howard Georgi and I suggested that there might be a common origin to all elementary particle forces. We proposed what have become known as Grand Unified Theories. Sadly, the simplest GUTs are wrong. They say that all matter is radioactive; that all atomic nuclei, even the proton, must decay with a known half-life. Although it may be true that diamonds are not forever, experimenters have shown them to live far longer than we anticipated. Does this mean that the forces of nature are not unified? Not necessarily. Particle theorists are loath to abandon this attractive hypothesis. My phenomenologically oriented colleagues have developed many imaginative and potentially unified theories. However, there is as yet hardly any empirical evidence supporting these elegant elaborations of the standard model. The super-conducting supercollider, once under construction in Texas, could have answered many of our questions, but it was unwisely aborted by Congress. All hope for progress in my discipline depends upon the successful completion of its less powerful sibling, Europe's Large Hadron Collider (LHC).

Meanwhile, a radically new and enormously popular approach to particle physics has arisen: superstring theory. Decades ago, the hope was expressed that the superstring could encompass all of fundamental physics: gravity as well as electromagnetism and the subnuclear forces, that it would yield a unique "theory of everything" describing the tiniest particles as well as the birth and evolution of the universe. Unfortunately, it has not worked out that way. In spite of what is regarded by its practitioners as enormous progress, the new theory makes not one explicit and verifiable prediction, nor has it recaptured the experimentally driven successes of the standard model. Perhaps someday the enterprise of particle physics will be divorced from the laboratory and become a purely cerebral activity. Not yet!

Nina Fedoroff

Plant biology, stress response of plants, transposons. Evan Pugh Professor of Biology, Willaman Professor of Life Sciences, Huck Institutes of Life Sciences, Pennsylvania State University.

I've never really thought about why I became a scientist. I guess I was too busy being one. When I was very small, I shared every little girl's ambition of being a ballerina. Later I took up music and decided to become a musician. So how did I end up a scientist? We know that our choices in matters of the heart find their reasons buried in ourselves, our genes, our childhood experiences. So, too, it must be in matters of intellect.

As a child, I loved reading, collecting information—and understanding. I remember writing a report about Russia when I was in eighth grade. It was an inch thick, typed and illustrated, and it had a gold cover to which I had pasted a black paper cutout of the Russian double-headed eagle. I loved doing it. It didn't seem odd to me at all. What surprised me was how much it seemed to surprise and please the teacher (though I seriously doubt that the other kids liked having it held up as an example).

In college, I loved literature and philosophy, psychology and political science. But even though some bits of philosophy and occasional literary insights seemed to transcend their time and place, most seemed quite stuck in the particulars of culture and history, at times just collections of people's opinions. Science seemed different. Writing a paper about experiments people had done was a bit like reading a mystery novel. It wasn't that people didn't have opinions and battles over who was right and wrong. It's just that each bit of research, each set of experiments, gave up another clue or two. However badly misread and misconstrued at first, they could in time be assembled to understand how something complicated worked by the workings of simpler rules.

I remember writing a paper on the amino acid code, the code that stores information in DNA, to be read out, then translated into the structure of a protein. DNA is a linear string of four chemicals called nucleotides. Proteins are linear strings of 20 different amino acids. Genetics told us that a particular DNA sequence uniquely determined the structure of one particular protein. But how? What was it that said, "Start here" and "put a lysine here"? Was it a string of nucleotides—like letters in a word? Was each word the same length? Were there spaces between the words? The answers came from experiments. But just how each experiment was designed to get an answer fascinated me. What I really liked was knowing that there was something I could do—an experiment—to test whether an idea was good or bad. An experiment is a reality check. You can argue all you want about the laws of physics, but when you throw an egg out of the window, gravity is inexorable and the results predictable. The experiment's outcome isn't a matter of personal opinion.

I also remember reading an essay in English class that was important in retrospect. I don't remember the author's name. But his point was that humans evolved on a scale between the largest and the smallest things in the universe—a realm in which causality works pretty well, though we now know that different laws rule in the realm of the very small and the very large. The author suggested that it should be no surprise that causality is comfortable for us. Causality probably works because survival and evolution guaranteed a good match between how our brain works and how the world we inhabit works. He cautioned us to be wary of causality for exactly that reason: because our brains "like" causality, we tend to attribute causal connections to all kinds of things. The only way of distinguishing real causes from fortuitous associations is through experiments, reality checks.

Then I started to *do* experiments, and I was hooked. Doing experiments turned out to be infinitely more absorbing than reading about them. Experiments are a way of discovering things no one has ever thought or seen before. Experiments can—and perhaps even should—be beautiful as well. I think I understood this through my work on transposons, often called jumping genes. They were discovered by Barbara McClintock, a great geneticist of the last century. Her experiments were elegant and spare—it was years before I stopped marveling at how much she had understood from the slimmest of clues. I remember a time when I finally went beyond what she could see just with genetics—I was so excited that I often couldn't sleep and sometimes forgot to eat.

So the answer is that for me science is not different from art, except in the one small, crucial detail that experiments speak their own truth, not ours. Science is very much a human creation, else our understanding of the world would never change. The inventive scientist plays with the observations and ideas, taking them apart and putting them together in different ways, adding new observations until suddenly a new pattern, a new reality emerges, one that itself must then be severely questioned with experiments.

Maclyn McCarty

Genetic material of cells. Professor Emeritus, Rockefeller University.

My mother said she could remember that at age 10 I read some scientific material that had been sent to my father, three or four books on things going on in biology. She also remembered that at age 10 I said I was going to be a doctor and do research. I never changed my mind. My high school in Kenosha, Wisconsin, had no German-language classes, so I went across town after school to get tutoring in German from a Lutheran minister. I had read that Johns Hopkins Medical School required a reading knowledge of French and German. It's not easy to say why I chose Stanford University, but that had to do with something I had read, too. My biochemistry professor there wanted me to stay and attend Stanford Medical School, but it had to be Hopkins because, again, of what I had read, in this case about the work there in medical research.

Upon graduation from medical school, I stayed on at Hopkins as a pediatric intern and resident with Dr. Edwards A. Park, who encouraged me to pursue a research career and helped me get a postdoctoral position with William S. Tillett at NYU. It was Tillett who suggested I apply for a National Research Council Fellowship for financial support and then contacted Dr. Oswald Avery when the NRC awarded me the fellowship, with the suggestion that I go to Rockefeller for further training. Avery's laboratory at the Rockefeller had been working for a long time with an organism called pneumococcus, the cause of pneumonia. Fred Griffin in England reported on a phenomenon he discovered, that by putting the bacteria into mice, the organism was transforming from one type to another. Avery picked up on this problem to try to find out why.

As time went on, it became clear that the transformation involved a change in one type of pneumococcus that was induced by a substance extracted from another kind. This change was predictable and permanent, being transmitted from generation to generation. It had all the earmarks of what we would call today the transfer of genetic information.

The varied types of pneumococcus were all very similar, but it turned out that they were different in their surface, called a capsule. Without capsules, the bacteria are not infectious. What we did was extract cell-free material from one type of bacteria and mix it with living bacteria of another type lacking capsules. The second type would then produce capsules of the first type. Gradually a number of pieces of evidence emerged to show that the transforming agent was DNA. It was the first experimental evidence that the genetic material in cells is DNA. In 1944, Oswald Avery, Colin MacLeod and I published this finding in a seminal paper in the *Journal of Experimental Medicine*, although the paper was not widely acknowledged at the time. Nonetheless, our discovery played a pivotal role, not only in the identification of DNA's structure by Watson and Crick nine years later, but it was also critical in the initial foundation of the fields of molecular biology and genetic engineering.

I have often said and written that the continued excitement in the research of those early days could hardly be duplicated again. Science today has changed in a way that's not good. There is too much involvement of fiscal gain in research. Researchers are dependent on financial gain and therefore work more for that than for what the research might do, what the value of it is for the cure of disease or for a variety of medical problems. As soon as you started getting things patented, you got involved heavily with the financial side. You can tell what was different in the early days because—for example, at Rockefeller—a whole slew of things were first turned up and not patented at all. The researchers weren't even thinking about getting the discoveries protected. People in medical science today, by and large, are much more oriented toward the dollar side of their work than when I started out.

Paul Greengard

Nerve-cell communication. Nobel Prize in Physiology or Medicine, 2000, for discoveries concerning signal transduction in the nervous system.
Vincent Astor Professor, Rockefeller University.

I went into science because I think better than I do anything else. As a child, I was always the last to be picked on a sports team. The only sport I ever succeeded at was in a Boy Scout jamboree where I came in first in the state of New York for potato sack racing. All the athletic frustrations of my childhood were eradicated in that one glorious afternoon. I then went back to thinking and the world of ideas.

My mother died in childbirth. She was, I have been told, a very brilliant person. I've noticed a great deal of discrimination against women in academia, and my wife and I decided to establish an annual prize in biology and medicine to be given to women. The Rockefeller University administers this prize. We have named the prize after my mother. The idea is to help women achieve a more equal status in science than they otherwise would. It's called the Pearl Meister Greengard Prize.

I was very good in mathematics and theoretical physics. The GI bill put me through college, and then I wanted to get support for graduate school. At that time, the only support available in the area of physics for graduate studies was in the form of fellowships from the Atomic Energy Commission. This was just three years after the atomic bombs were dropped on Hiroshima and Nagasaki, and I didn't want to contribute to a field which might then be exploited to make more potent weapons. I went into the biophysics of the nervous system, honing my skills in terms of understanding both the biochemical and the electrical properties of nerve cells. The study of the biochemical basis of nerve cell function was virtually nonexistent at that time.

As a result of our work, it is now clear that there are two basic types of synaptic transmission, termed fast and slow. Fast transmission from one cell to another across a structure called the synapse takes about one-thousandth of a second. One cell releases a chemical called a neurotransmitter, which then activates a second cell, the target cell. This fast synaptic transmission was the only kind that was partially understood at the time. The reason it's so fast is that the neurotransmitter binds to a protein on the surface of the target cell and causes a change in the protein's structure so that ions flow through it. There's only one protein involved.

My work contributed to our understanding of what we now call slow synaptic transmission. What we call fast transmitters today are primarily glutamate, which is an excitatory fast transmitter, and a compound called GABA, which is an inhibitory fast transmitter. I was interested in how other neurotransmitters work. We now know that there are dozens and dozens of other neurotransmitters that affect nerve cells. We elucidated the principles by which those compounds produce their effects on their target cells. They produce them through what we call slow synaptic transmission. The slow synaptic transmission produces effects not by changing the structure of a single protein, as in the case of fast transmission, but by a whole cascade of biochemical steps of unbelievable complexity. The practical importance of this knowledge is that it has provided a lot of new targets for the development of new drugs to treat such diseases as Parkinson's, schizophrenia, Alzheimer's, depression, anxiety, and drug abuse.

Drugs have been developed that mimic or antagonize the actions of neurotransmitters on the surface of cells, depending on whether there is too little or too much of a desired signaling activity. For example, Parkinson's is treated by increasing dopamine signaling. Schizophrenia is treated quite effectively by compounds that decrease dopamine signaling. However, the situation is complex with schizophrenia. It's clear that the dysfunction in Parkinson's results from too little dopamine, so that you can treat it fairly effectively by giving dopamine. You can also treat it by giving other compounds that antagonize those pathways in the cell that oppose dopamine in order to make the dopamine more effective. The logic behind using a dopamine precursor, levodopa, to treat Parkinson's is very clear. Since we don't yet understand the cause of schizophrenia, the logic for treatment is less clear. A lot of people, including myself, would love to establish a firm logic for treating schizophrenia.

In terms of the brain, you can in a crude way think of the human brain as a computer. Furthermore, you can think of fast transmission as the hardware of the brain and slow transmission as the software that directs the hardware whether or not to communicate. The fast transmission tells the cell to fire or not fire, whereas the slow transmission is much more subtle.

I continue to be interested in and to work on trying to understand the basis for various neurological and psychiatric disorders and the mechanism of action of various therapeutic drugs and drugs of abuse. Those are the things we are doing now.

Alan Guth

Particle physics. Victor F. Weisskopf Professor of Physics, Massachusetts Institute of Technology.

As a child, I was perhaps more of an engineer than a scientist, but at that stage I was not even aware of the distinction. A cousin of mine won't let me forget that one of my earliest projects was to try to grow a money tree. That project failed. Later, I think when I was in the seventh or eighth grade, I tried to build a computer that would play tic-tac-toe. It was a box containing batteries, switches, flashlight bulbs, and a maze of wires. The human player would register moves by turning a switch. The switch would then close the circuit to light a bulb indicating the computer's move—or at least so I planned. I'm not sure that I ever got it to work for all combinations of moves, but it was certainly more successful than the money tree project.

My interest in theoretical physics began in high school, and I pursued the subject as an undergraduate at MIT. I went on at MIT to get a Ph.D. in theoretical particle physics, the study of the elementary building blocks that make up atoms and all of the universe. It was not until eight years after completing my Ph.D., however, that I was dragged into cosmology by a fellow postdoc at Cornell, Henry Tye. He persuaded me to join him in trying to calculate the production in the early universe of a hypothetical particle called a magnetic monopole. We found that these particles could easily be produced in large numbers, so we had to work hard to find assumptions that could avoid the prediction of a universe filled with monopoles. Continuing on my own when Henry left for a visit to China, I found that these ideas led naturally to a new cosmological theory, now called the inflationary universe. I soon realized that the theory had a crucial flaw, but I published it anyway, explaining both its successes and its failure. The failure was soon overcome by a modified form of the theory, discovered by Andrei Linde in the Soviet Union and independently by Andreas Albrecht and Paul Steinhardt in the United States.

Inflation is very exciting, because the repulsive gravity it describes can be the explanation of the driving force behind the Big Bang expansion, and it turns out that the theory can even explain the creation of essentially all the matter and energy in the universe. (Inflation is not, however, a theory of the ultimate origin of the universe, since one needs a small amount of preexisting matter for inflation to start.) Inflation can explain a number of features of the universe that otherwise seem mysterious. First, there is the large-scale uniformity of the universe, the fact that the universe looks about the same in all directions. In conventional Big Bang cosmology, this fact is hard to understand, since according to this theory the distant galaxies are so far away from each other that they have never had any effective contact. In the inflationary theory, however, the whole universe could have become very uniform while it was tiny, and afterward the repulsive gravity of inflation stretched it to an enormous size. Second, cosmologists have never understood what determined the initial expansion rate of the universe, which has to have been very finely tuned to produce a universe like the one we see. If the initial expansion rate had been just a tiny bit lower, even by just one-billionth of a percent, the universe would have long ago collapsed under the force of its own gravity. If it had been just a tiny bit larger, the universe would have flown apart so fast that there would have been no time to form galaxies, stars, or planets. It turns out, however, that the repulsive gravity of inflation drives the universe at just the right rate to avoid either of these cosmic catastrophes.

Finally, cosmologists have for many years assumed that the matter in the early universe was almost uniform, except for small ripples that were needed to explain how the matter later clumped to form galaxies. We can now actually observe these ripples as tiny nonuniformities in the cosmic background radiation, a bath of microwave radiation that seems to permeate the universe, and which we believe is the afterglow of the heat of the Big Bang. Nonetheless, before the theory of inflation, there was never any understanding of where these ripples came from. In inflationary models, these ripples are traced directly to the intrinsic unpredictability of quantum physics, which leads to the conclusion that inflation ended at slightly different times in different places. Inflation makes predictions about the statistical properties of these ripples, and it has been very exciting over the past 10 years to follow the increasingly precise measurements of these ripples. So far, the measurements seem to be completely in accord with the predictions of inflation. I continue to be fascinated with the fact that calculations can actually predict real things about the real world.

The excitement in cosmology continues, because inflation is by no means the end of the story. Inflation is not really a theory, but rather a class of theories, with no particular version that seems compelling. While theorists like me are exploring the possibilities, new data are appearing at a remarkable rate. We clearly still have a lot to learn about how the universe began.

MASSACHUSETTS INSTITUTE OF TECHNOLOGY PHYSICS COLLOQUIUM
Inflation and False Vacuum Bubbles
MASSACHUSETTS INSTITUTE OF TECHNOLOGY PHYSICS COLLOQUIUM
LOCALIZED INFLATION, BLACK HOLES & CHILD UNIVERSES
MASSACHUSETTS INSTITUTE OF TECHNOLOGY PHYSICS COLLOQUIUM
THE NEW INFLATIONARY UNIVERSE
72T

David Helfand

Large-scale structure of the universe. Professor of Astronomy, Columbia University.

It is convenient, I have found, to develop a neatly packaged mythology about one's life to present to new acquaintances, to colleagues, even at times to oneself. While perhaps not fully consistent with the philosophy expected from a late-1960s coming-of-age, this approach does comport with my much maligned motto "The examined life is not worth living" (much maligned, that is, by my artist spouse).

An essential part of my myth involves my career choice while at Amherst College. Although as politically active and appropriately left-wing as most of my compatriots in the class of '72, I had a great deal of trouble taking seriously the mantra of "relevancy" which droned on in the background of every conversation about career choice in those halcyon days; assertions concerning the social relevance of investment banking and plastic surgery always seemed to me a trifled strained. And so with characteristic irreverence, my mythology goes, I chose the most irrelevant career I could think of—understanding the origin and evolution of the universe. I frankly can't remember (too much living and not enough examining, I suppose) whether I effected this stance prior to graduation or not. But for one whose few surviving strands of hair remain resolutely pulled back in a ponytail flopped over his Ivy League suit collar and who recently threw a party to celebrate 31 years and eight months (exactly one billion seconds) of not shaving, my commitment to a very un-60s irrelevancy is usually enough to get a conversation started.

In fact, I became an astrophysicist one evening in the backseat of a car in Arizona during my sophomore year in college. Owing to the remarkable generosity of one of my professors, a few classmates from the previous semester's astronomy course and I were spending January visiting places where astronomy is actually done; that evening we were being taken on an observing run by one of the major figures of 20th-century astronomy, Professor Bart Bok. An apparently gruff, but wonderfully warm, Dutchman, Professor Bok asked from behind the wheel whether any of us were actually considering careers in astronomy. Appreciating the warm Arizona evening (it was 20 below back in Amherst), I volunteered that I was (despite still being registered as a drama major). His first question: "If you were called to testify before Congress to discuss the reason they should fund astronomy, what would you say?" Somewhere, I had already picked up what I thought to be the politically astute response, and I started into a disquisition on the technical spin-offs, the value for science education, and so on. He abruptly interrupted, "No, that is *not* why we do astronomy. We do it as part of an ageless quest for perspective on our place in the universe, an aesthetic pursuit which distinguishes us as human. You should always argue for support on the same grounds you would use to make the case for symphony orchestras and poets." And since then, I always have.

It happens to be the case that astronomy is also great fun. I left Amherst 27 years ago and, having discovered the center of the universe (Manhattan), I now seek to explore its edges. This has sometimes meant frustrating nights on a mountaintop in Chile listening to the rain drumming on the shuttered telescope dome. It has included scrambling for support for my students, sitting through mind-numbing NASA committee meetings, and engaging (a trifle more often than absolutely necessary, I must confess) in those titanic academic battles in which the warriors are so vicious because the stakes are so small. But each time a few photons of light, having traveled uninterrupted for 11 billion years, are captured by my telescope and an image of a distant corner of the universe never before glimpsed scrolls up on my computer screen, I revel in my decision to be irrelevant.

Sir Andrew Huxley

Cell biology. Nobel Prize in Physiology or Medicine, 1963, for discoveries concerning the ionic mechanisms involved in excitation and inhibition in the peripheral and central portions of the nerve cell membrane. Cambridge University.

My boyhood interests were mostly of a mechanical kind. I built quite complicated models with Meccano sets and also put together clockwork trains and did a fair amount of woodwork as well. At Westminster School, I started in classics. When it became clear that my real interest was in mechanical things, I switched to the modern side and was extremely well taught in physics. Then, after five years at school, I came up to Trinity College, Cambridge, of which I am now a fellow. My original intention was to specialize in physics and probably become an engineer, but my interest gradually turned to physiology. One of the people I met in college was Alan Hodgkin. A few years older than I, he had become a junior research fellow in Trinity, and he was already distinguished in experimental work on nerves. In the summer of 1939, Hodgkin went to Plymouth, where there is a marine biology laboratory, which is the only place in England where it is possible to get squids for experimental work. Squids have a pair of nerve fibers which are enormously larger than ordinary nerve fibers—about half a millimeter across. That means that you can do many sorts of experiments on them that are not practicable on what you might call ordinary nerve fibers. In particular, you can push things down inside them.

I joined Hodgkin, and we did push down an electrode, a wire for recording electrical changes from inside of the fiber. As expected, the inside was about a twentieth of a volt negative relative to the fluid outside. The message that goes along nerve fibers is a sequence of electric pulses which are generated as they go along, so they maintain themselves. In that respect it's like a train of gunpowder burning. The theory current at the time was that when the inside of the fiber became a little less negative than it is at rest, the membrane separating it from the external fluid became freely permeable to all ions, in particular to cations (positively charged ions like sodium and potassium), and it would act like a short circuit so that the electric potential of the inside of the fiber would rise to very close to the external potential. What Hodgkin and I found was that it did not merely approach the external potential, but it overshot and went something like 50 millivolts positive in the inside during this short-lived impulse, which lasts about a thousandth of a second.

During the war, we both turned to other things, but I visited Hodgkin a number of times and we wrote a paper in which we produced no fewer than four possible explanations for this overshoot. All of them turned out to be wrong. With hindsight, I certainly feel very stupid not to have jumped to the right conclusion straightaway, which is that this great increase of permeability in the membrane is not indiscriminate but is highly specific for sodium ions. Sodium ions are much more concentrated in the external fluids than they are inside, so when the membrane becomes highly permeable to sodium, sodium ions enter, carrying their positive charge, and that is what makes it go positive. As often happens in science, when you know the answer, it seems terribly obvious. We did much further work on this subject, leading eventually to the Nobel Prize that we received jointly in 1963. The references in the citation to inhibition and to the "central portions of the nervous system" (i.e., the brain and spinal cord) apply to the work of Sir John Eccles, who shared the prize with Hodgkin and me.

I think that in anything that matters, scientists look for actual evidence. Many scientists, including myself, follow my grandfather, Thomas H. Huxley, in taking an "agnostic" position about any kind of supernatural being or persistence of a separate soul. He coined that word to convey that religion was a field about which it was not possible to have any definite knowledge. It is definitely not an atheist position, but it is a position of not knowing and being aware that we don't know anything reliable about any of these things: my position is an agnostic position. I don't actively believe, and I don't deny the existence of a God. Sir John Eccles was openly a Catholic and believed in the separate existence of a soul. He wrote books about this and put forward a hypothesis about what the elements in the brain were that were influenced by the mind, but he really had no evidence. This was a hypothesis, and it's a great help in starting on any scientific work to have a hypothesis to test. Books about the relation between mind and matter have been written by other very distinguished scientists, but as far as I know, no one has reached conclusions that are at all firm. There certainly are a number of excellent and distinguished scientists who, like Eccles, are believers in some religion. We're all in a state of uncertainty.

David Hubel

Mammalian visual system. Nobel Prize in Physiology or Medicine, 1981, for discoveries concerning information processing in the visual system. John Franklin Enders University Professor of Neurobiology, Harvard University.

I was born in 1926 in Windsor, Ontario, Canada, and grew up in Montreal, where my father worked as a chemist. My parents, who were American, registered me at birth as an American citizen, so as a Canadian by birth and an American by derivation I grew up a dual citizen, with all the mixed loyalties that implies.

English-speaking children in our Montreal suburb went to a "Protestant" school—"Protestant" meaning non-Catholic. The teachers were all Protestants because, in Quebec, Catholics were not allowed to teach in "Protestant" schools. This had the unfortunate effect that French was taught either by English-speaking teachers or by Huguenots imported from France. No effort was made to teach us to speak or understand French as spoken with a French-Canadian accent, with the result that few of us could converse in French.

As a child, I spent much time experimenting with chemicals. Another hobby was electronics, and I built several radios. Long afterward, when I was 67, I passed five sets of ham radio tests, learned Morse code, and took on still another hobby.

Music has always been a major interest. I took piano lessons from age five onward. Perhaps luckily, I was only moderately talented and was a hopelessly bad sight reader; had it been otherwise, I might have been tempted to go on with music as a career. I have wondered if my bad sight-reading, like my slow speed at ordinary reading, may be related to my being left-handed.

At college age, I went to McGill College in Montreal. I had developed a love for mathematics and physics and enrolled in a combined honors course in those subjects. Mathematics was excellent, and although it hasn't turned out to be directly useful in my later work, it taught me the dangers of relying on reasoning that involves more than a very few steps.

At the end of my final year in college, I had to decide what to do next. I did not want to be a mathematician and doubtless lacked the virtuosity. Attending a physics conference in Montreal in 1943 showed me how inadequate my physics training had been, but for want of an alternative I applied and was accepted to graduate school in physics at McGill. Almost on a whim, I also applied to medical school, and to my horror was accepted there too—so now I was faced with my first big career decision. I had no background or experience in medicine, but I suppose I had some vague ideas about applying physics to medical research. I chose medical school.

After graduation, I did the usual internship and a two-year residency in neurology. By 1954, I was eager to start doing research and turned my attention toward neurophysiology and so—although I could not have imagined it at the time—set out on the path that led to the award of the Nobel Prize to me and my longtime collaborator, Torsten Wiesel. Our work was on the responses of cells in the mammalian visual system. Notwithstanding the great progress that we and others have made in this field, the present state of knowledge is only the beginning of an effort to understand the physiological basis of perception.

Richard Darwin Keynes

Transmission of nerve impulses.
Author, formerly Professor of Physiology, University of Cambridge.

Since my mother's maiden name was, and my own middle name is, Darwin, you might suppose that this explains why I inevitably became a scientist. But without denying the possible existence of a specific gene for science, I should properly acknowledge my ultimate arrival on the pages of this book to the encouragement, from my boyhood onward, of the 1922 Nobel laureate A. V. Hill, husband of my father's sister, to follow his example by becoming a physiologist. And when on my entry in 1938 to Trinity College, Cambridge, I found myself being taught by Alan Hodgkin; the wisdom of A.V.'s advice was amply confirmed.

During a wartime interruption to my education lasting for five and a half years, while I served as a civilian scientist working for the British navy first on the design of sonar systems for submarine detection and then on the design of gunnery radar sets, I acquired a very useful knowledge of electronics. In 1945, I returned to Cambridge to complete my degree course in physiology, again under the tutelage of Alan Hodgkin. This was an extraordinarily favorable moment for embarking on a research career in neurophysiology, because a wide choice of fundamentally important problems in this area were then ready to be tackled by the new techniques that had become available during the war. So while Alan Hodgkin and Andrew Huxley followed up the crucial breakthrough that they had made in 1939 by recording the internal electric potential in a squid giant nerve fiber, which led to their classical studies by electrical techniques of the mode of conduction of the nervous impulse, I set out to examine other aspects of the problem with the aid of radioactive isotopes of sodium and potassium. To begin with, I had to manufacture the isotopes for myself, but the prewar cyclotron at the Cavendish Laboratory had fortunately been brought back into service, and it was not too hard to produce adequate samples of the short-lived ^{24}Na and ^{42}K (isotopes of sodium and potassium, respectively, Na and K being the chemical symbols for those elements) by deuteron bombardment of suitable targets. After checking that my measurements of the movements of labeled Na+ and K+ ions during a single nerve impulse agreed satisfactorily with the electrical ones, I joined forces with Alan Hodgkin to work every autumn for ten years on squid axons, in the course of which we examined the way in which the Na+ ions that had leaked in during the impulse were pumped out afterward, this being a process that could be examined only by the isotope technique. Then, until in 1993 my fingers became too stiff to dissect even giant nerve fibers, I continued to spend a month or two almost every autumn doing experiments mainly on the properties of the sodium gating currents of squid axons at the Plymouth Marine Laboratory, or in the last seven years at the Roscoff Laboratory in Brittany.

There is not infrequently in a scientist's career a lucky break that influences the later development of his work in a beneficial and wholly unforeseen way. This befell me in 1951 when Professor Carlos Chagas Filho invited me to go out for three months to the Instituto de Biofisica in Rio de Janeiro as visiting reader. I was thus enabled to use the intracellular microelectrodes just developed in Cambridge to show for the first time exactly how the electric eel generates such large pulses of electricity, which was a valuable technical advance. But of greater significance to me in the long run was the satisfaction that I continued afterward to find in meeting and collaborating with scientists from Brazil and other parts of South America. This later involved me deeply in the work of the International Council of Scientific Unions (ICSU) and, most particularly, in service as chairman of the ICSU–Unesco International Biosciences Networks, whose object was to encourage the development of biological research in the countries of the Third World. And finally it was in Buenos Aires in 1968 that by pure good chance I came across some little-known pictures painted by Conrad Martens on board HMS *Beagle* in 1833, and belatedly found myself a new interest in the researches on geology and biology carried out by my great-grandfather on his famous voyage. After first publishing a book entitled *The Beagle Record*, in which I catalogued the Martens drawings, I produced a new edition of Charles Darwin's *Beagle Diary*, and then transcribed *Charles Darwin's Biology Notes & Specimen Lists* from H.M.S. *Beagle*, which had not previously been published. Most recently, I have written a book entitled *Fossils, Finches and Fuegians*, a retelling of the whole story for the general reader. To have a fresh occupation as an historian of science to cap 45 years of experimental work that included a fair share of teaching and administration has been very rewarding.

Ruth Patrick

Freshwater ecology. Francis Boyer Chair of Limnology, Philadelphia Academy of Sciences. Adjunct Professor, University of Pennsylvania.

When I was a very small child, my father took me on walks in the woods every Sunday afternoon in Kansas City, where we lived. My older sister and I each had our little basket, and Daddy would have a fishing pole like Tom Sawyer. We would collect bugs, worms, mushrooms, flowers, grasshoppers, everything that interested us. We would bring them home and have milk and crackers, and then Father would open the doors to the library. We lived in an old-fashioned house with great big wooden doors that Father would roll back. We'd go in and he'd open his rolltop desk, where he kept four microscopes. He'd pull out the one that was appropriate for what he wanted to show us. If we looked at water, which fascinated me, he would make a glass slide mount and put it under the microscope. I'd crawl up on his knees and peer through the microscope. It entranced me to see such entirely different worlds with the naked eye, very different from what you see with trees and flowers. I became fascinated.

Then he gave me the microscope he'd been given when he was a little boy. I'd lie on the floor, particularly in the dining room where there was a north window, and make mounts of all kinds of things and peer through the microscope. I loved the microscope for its intrinsic self but also because I admired my father so much. Every night before he went to bed, he would look at something through his microscope. This was long before the time of radio and television, and many, many people who had leisure time, particularly men, made fancy mounts and sent them around to their friends to see who could identify them. In the 1880s into the early 1900s, there was a rivalry between the men as to who could see the most.

My father continued to interest me in science, and then in school my science teacher, seeing my interest, promoted it. I took appropriate courses in high school and college to enhance my knowledge. In college, I majored in biology and the precursors of limnology. I took courses in microscopic analysis, algae, vertebrate zoology, and protozoology. I went to a very small school in South Carolina called Coker College. Coker College was the best thing that ever happened to me because I was shy, and in a small school I was amazed to have leadership roles. That developed my self-confidence.

Every summer, my father would send me to Woods Hole or Cold Spring Harbor, where I met very excellent people. That's where I met Dr. Ivey F. Lewis, who was a specialist in red algae and under whom I got my Ph.D. at the University of Virginia.

I continued the study of diatoms. I was the first to model the structure of the community of algae to show the effects of pollution. And in focusing on diatoms, which are algae with distinctly (and beautifully) shaped silica shells, I discovered that the presence of particular species in a body of water indicated the chemical condition of the water. And the chemistry of the water indicated the pollution of the water draining the watershed. Here was a window to a much broader objective: understanding the environmental impact of human activities by looking at the biology of the ecosystem in which they are taking place.

I have also studied climate and water conditions in the archaeological record as they are reflected in the shells of ancient diatoms. For example, I studied diatoms from Clovis, New Mexico. There was evidence that a mammoth-like animal had disappeared from the region and that a small horse-like animal later appeared in the same region, but no one knew why the change took place. By examining the diatoms associated with the remains of these animals, I was able to say that the mammoth bones were associated with very cold freshwater and that the diatoms associated with the horse-like animal were warm-water species. I postulated that the extinction of the mammoth was due to a change in world temperature. This theory is now accepted.

François Jacob

Cell biology. Nobel Prize in Physiology or Medicine, 1965, for discoveries concerning genetic control of enzyme and virus synthesis. Institût Pasteur, Paris.

When I came back from World War II, I was heavily wounded. I could not become a surgeon as I had wished. I turned to science. Although I knew very little, I understood that something was going to happen at the interface of bacteria, biochemistry, and genetics. That is why I began to work on the genetics of bacteria and the cellular machinery.

It was the work of a lifetime, involving countless experiments and many failures. And successes. With the publication in the *Journal of Molecular Biology* of "Genetic Regulatory Mechanisms in the Synthesis of Proteins," Jacques Monod and I were able to prove the existence of messenger RNA and its function in cells. The work showed for the first time how a gene functions; how it sends a continuous stream of information toward the cytoplasm (the part of a cell surrounding the nucleus), rather like a faucet whose flow can be regulated according to the requirements of the cell as a function of signals from the environment. It proposed a model to explain one of the oldest problems in biology: in organisms made up of millions, even billions of cells, every cell possesses a complete set of genes; how, then, is it that all the genes do not function in the same way in all tissues? That the nerve cells do not use the same genes as the muscle cells or the liver cells? In short, this article presented a new view of the genetic landscape.

I have felt that life is a constant race with time. It took me a long while to realize that this drive toward tomorrow has an advantage in at least one domain: in research. Late, very late, I discovered the true nature of science, of how it proceeds, of the men and women who do it. I came to understand that, contrary to what I had believed, the march of science does not consist of inevitable conquests, or advance along the royal road of human reason, or result necessarily and inevitably from conclusive observation dictated by experiment and argumentation. I found in science a mode of playfulness and imagination, of obsessions and fixed ideas. To my surprise, those who achieved the unexpected and invented the possible were not simply men of learning and method. More than anything else, they possessed extraordinary minds, enjoyed the difficult, and often were creatures of amazing vision. Those in the front ranks displayed exotic blends of passion and indifference, of rigor and whimsy, of naïveté and the will to power, in a triumph of individuality.

Eric Kandel

Molecular mechanisms of memory storage. Nobel Prize in Physiology or Medicine, 2000, for discoveries concerning signal transduction in the nervous system. University Professor, Departments of Physiology, Psychiatry, and Biochemistry, Columbia University. Senior Investigator, Howard Hughes Medical Institute.

I was born in Vienna, Austria, and immigrated to the United States in April of 1939, thirteen months after Hitler annexed Austria and released in Vienna a violent outburst of anti-Semitism. We settled in Brooklyn, where I first completed elementary school and then attended Erasmus Hall High School. Upon graduating from Erasmus, I went to Harvard College in Cambridge, Massachusetts, where I studied contemporary history and literature. I wrote my honors dissertation on the attitudes toward National Socialism of three German writers, each of whom represented a different point on the political spectrum. I tried to understand what happened in Germany and Austria in the 1930s. How could a people be so cultured and civilized at one moment in history and so cruel and destructive the next?

As I was completing college, I became fascinated with psychoanalysis and thought it to be a more original and profound approach to understanding motivation than intellectual history. I therefore went to medical school at New York University, with the idea of becoming a psychoanalyst. While in medical school I took an elective period in brain science, thinking that even a psychoanalyst should know something about the brain. To my surprise, I found that I much enjoyed doing research. Based on my experience in his laboratory, my mentor Harry Grundfest recommended me to the NIH for a postdoctoral fellowship in neurophysiology as an alternative to the then existing doctor's draft. I spent three years, 1957 to 1960, at the NIH focusing on the biology of memory, an issue I thought central to psychoanalysis. I worked on the biology of the hippocampus, a structure in the mammalian brain that had just been shown to be critical for memory storage of complex events. I left the NIH in 1960 and began residency training in psychiatry at the Massachusetts Mental Health Center of the Harvard Medical School. Upon completing psychiatric training in 1962, I returned to basic science.

I now decided not to continue to study memory in its most complex form in the hippocampus but to take a more radical, reductionist approach based on the idea that memory storage is likely to have a general, conserved, cellular, and molecular mechanism and this might be approached most effectively in a simple animal and in a simple learning task. I therefore searched for an appropriate animal to study learning and memory and focused on the marine snail *Aplysia*, which has the advantage that it has only 20,000 nerve cells rather than the billions of nerve cells of the mammalian brain. In the snail, I was able to delineate a very simple reflex behavior controlled by fewer than 100 cells and to pinpoint changes that occurred with learning and memory storage. These experiments provided the initial evidence that learning involved alterations in synaptic strength that are maintained as memory, and that long-term memory takes on its special enduring character by altering gene expression and giving rise to the growth of new synaptic connections. In the year 2000, I was privileged to receive the Nobel Prize in Physiology or Medicine for this work, which I shared with Paul Greengard and Arvid Carlsson.

In retrospect it seems a very long way for me from Vienna to Stockholm. My timely departure from Vienna made for a remarkably fortunate life in the United States. The freedom that I have experienced in America and in its academic institutions made Stockholm possible for me, as it has for so many others. Having been trained in the humanities, where one learns early how depressing life really can be, I am delighted I switched to biology, where a delusional optimism still abounds. I continue to explore the science in which I worked almost like a child with naïve joy and amazement. I sense myself particularly privileged to be in the field of neurobiology of the mind, an area that, unlike my first love, psychoanalysis, has grown magnificently in the last 50 years. Looking back on these last 50 years, I find myself surprised to be doing what I am. I entered Harvard to become a historian and left to become a psychoanalyst. But I am glad that somehow or other I followed my instincts rather than my early training and ended up doing something that I have immensely enjoyed.

Cynthia Kenyon

Neuroscience. Regulation of aging.
Herbert Boyer Distinguished Professor of Biochemistry and Biophysics, University of California, San Francisco.

I was a little truth seeker as a child. I wanted more than anything to understand myself and also other people. I remember that my best friend stenciled for me "Seek Ye the Truth, and the Truth Shall Set You Free," and I hung it on my bedroom wall. My path to truth was winding. It started with Russian novels, then psychology courses, and eventually led to molecular biology after my mother gave me a copy of Jim Watson's wonderful *Molecular Biology of the Gene*. I also loved learning. I have always gotten a thrill, a kick, from learning new things. Plus, very important, from a young age I wanted to accomplish something truly great, something extraordinary, during my lifetime.

As a scientist, I first studied problems pioneered by others—for example, the question of how a fertilized egg develops into a complex, fully formed animal. During this time, I began to realize two very important things about biology: nothing, it seemed "just happened"; everything—from cell division to biological pattern formation—was subject to elaborate and clever modes of regulation. Second, these regulatory mechanisms were, to a remarkable extent, the same in all organisms. The study of development was fascinating, but I still wanted to find something completely new and unknown to study. That turned out to be aging. Everyone seemed to think that aging was something that just happened. We wear out, like old cars. But genes had to influence aging. After all, mice live 2 years but bats can live 50. Rats live 3 years; squirrels, 25! Also, how could a young girl go through puberty at age 12 and then menopause four decades later if the aging process weren't regulated in some way? I decided that there might really be something interesting here. Perhaps genes did regulate the aging process. Perhaps different organisms had different life spans because a universal regulatory "clock" was set to run at different speeds in different species.

We looked for genes affecting aging in a small, short-lived roundworm called *C. elegans*. We discovered that changing a single gene, called *daf-2*, doubled the worm's life span and kept it youthful and disease-free much longer than normal. The *daf-2* gene encoded a hormone receptor—that is, a protein that allows tissues to respond to a certain hormone. So aging *is* regulated, by hormones. We showed that the hormones affect aging by coordinating the activities of many downstream genes, including antioxidant, metabolic, and antimicrobial genes. Stimulated by our findings, others found that the same hormones control aging in flies and mice, suggesting that they will also turn out to control aging in people. Aging has puzzled and bedeviled mankind for centuries (just think of Shakespeare's sonnets), so our findings give me a deep sense of satisfaction and happiness. Maybe one day we will be able to take a pill that keeps us young and healthy much longer. I believe in my heart that this will happen.

Sir John Richard Krebs

Animal behavior. Chairman, UK Food Standards Agency. Royal Society Research Professor, Oxford University.

Why did I become a scientist? What does it mean to me?

One simple answer is that my father was a Nobel Prize–winning scientist. But although our parents are role models, most of us strive to shake off the shackles of parental influence. I tried to do this at high school, where I preferred languages to science and where my main hobby (apart from sport and dreaming about girls) was archaeology. But I didn't totally break free and ended up studying zoology at university.

Not that I regretted it for a moment, especially when a few years later I discovered that I could actually get paid for watching birds, another boyhood hobby. I was also lucky enough to have inspiring and charismatic teachers at university, and during holiday work experience in an institute in Germany.

I have spent much of my research career trying to understand the behavior of animals in their environment: for example, "Why do some animals live in groups whilst others are solitary?" "How efficient are animals at exploiting the food resources in their environment?" "How do animals respond to unpredictability and risk?"

My research has celebrated the diversity of living things and explanations for this diversity. In contrast, my father in his work elucidated a universal feature of all living cells—how food is broken down to form the energy that is essential for life. It was only near the end of his life, when I was an established scientist, that my father and I debated our complementary ways of studying living things and captured our thoughts in a short paper.

For me, science has many similarities to art. Both are creative, both aim to express the essence of the world about us or our inner selves, and in both a sense of aesthetic is central. Scientists talk of a "beautiful experiment" or "an elegant explanation" with a similar sense of aesthetic appreciation that we have for Mozart or Miró. The parallel doesn't go all the way, though. Science is different from other forms of creative endeavor in the way that observations and theories accrete through time, so that we can say with confidence that our understanding of, say, genetics is deeper now than it was a hundred years ago. We cannot say with equal confidence that, for instance, Rothko represents an advance on Rodin.

When we learn science at high school, we tend to learn it as a set of facts—Newton's laws, the classification of animals, and the periodic table. So people often expect scientists to "come up with the answers" when there is a problem. But often science is incomplete: scientists do not have the definitive answer, although they may have a way of finding out. In recent years, when I have been responsible for food safety in the UK government, I have often found myself having to say to the media that scientists cannot calculate for certain the size of a particular risk, such as BSE (bovine spongiform encephalopathy, or mad cow disease).

Perhaps one of the most satisfying things about science is its shared purpose. As a scientist, you are truly part of a global endeavor. You enjoy the intimacy of exchanging knowledge and ideas that transcend cultural, religious, and language barriers.

Leon Lederman

Particle physics. Nobel Prize in Physics, 1988, for the neutrino beam method and the demonstration of the doublet structure of the leptons through the discovery of the muon neutrino. Director Emeritus, Fermi National Accelerator Laboratory. Pritzker Professor of Science, Illinois Institute of Technology.

My first recollection of science was the book written by Einstein with Leopold Infeld on the meaning of relativity. In that book, read when I was ten and recovering from the measles, the authors compare science to a detective mystery. After a number of clues, the detective provides a complete accounting for all the clues, eliminating all but one suspect. The connection here is the use of rational thinking, guided by the clues provided by observation. In my science, the observational techniques have developed to giant and expensive "atom smashers"—i.e., particle accelerators, which provide us with probes that allow us to explore the structure of matter down into the atoms (a thousand billion atoms can fit on the period that ends this sentence) and deeply into the nucleus, itself a trillion times smaller than the atom.

My high school experience was a medley of teachers and friends, each stimulating my curiosity. A joy of learning grew as I began to read about science and about scientists. I recall the excitement of reading about Niels Bohr's 1913 discovery of how the atom's structure can explain the beautiful spectra of colors that can be precisely measured with the aid of a prism and a telescope.

College continued this process of knowing, of knowing how science works, how even a most elegant and beautiful theory can be destroyed by an ugly fact, but then the insights and crystalline logic of an Einstein or a Feynman can reveal form and breathtaking beauty in nature. How, I wondered, can minds be so powerful and nature so resourceful? I "grew up" in graduate school at Columbia, using what was, in 1950, the world's most powerful cyclotron, enabling one to study the properties of new particles, witnessed in the catastrophic collisions produced by the machine and studied in an elegant device, a cloud chamber. This 1930s technology evolved to enable my generation of young physicists to reveal a surprising complexity in the nature of the subnuclear microworld.

The accelerators, giant microscopes, produce the collisions, and the particle detectors that observe and measure are like the air I breathe. They are the instruments of humanity's reach into the depth of inner space. The next three decades, based upon ever increasingly powerful machines, have given us a new picture of mankind's conceptual grasp of the structure of matter and energy, of space and time. Today, this understanding of inner space, even though far from complete, turns out to be an essential half of the answer to the question "How does the universe work?" The other half comes from outer space, the study of stars, galaxies, black holes, quasars, and a working model of the very creation and evolution of the universe from its inception in the explosion of a hot soup of our particles to the world we see on a dark winter's night far from the city lights. This union of inner space and outer space is perhaps the most stunning discovery of our times. It is a story in progress.

That this story is so little known and so little appreciated speaks to the failure of education in this nation. That mankind has always been driven to understand the world in which we are embedded is part of the tragedy. Perhaps a more salient and more urgent part is the failure of our society, of all people, to grasp the qualities that make science (and its associated technologies) so successful. It is true that the comforts and the profits that are suggested by cell phones and fax machines, by computers and the Internet, by jet planes and television—none of this would be possible were it not for the knowledge which came out of our efforts to know how the universe works.

However, what is a more egregious failure is to be unable to communicate the aesthetics and the qualities of mind that are derived from all of this and that are available to all—in particular, to the graduates of the high schools of America. Our high schools need a profound revolution in how they teach science, trapped in curricula that are one hundred years old and out of date. We need to enter the 21st century to teach the revolutions of the past century and, with these, the attributes of curiosity, of skepticism, of integrity, of openness of mind, of the value of diversity, of ethical and moral sides. Without these, there would be no science, and without these, a democratic society may not survive the darkness of rigid belief systems and the greed and fear that so infects human nature.

Eric Lander

Genomics and genetics. Director, Eli and Edythe L. Broad Institute and Harvard Professor of Biology, Massachusetts Institute of Technology. Professor of Systems Biology, Harvard Medical School. Member, Whitehead Institute for Biomedical Research.

The secret of biological evolution lies in the random exploration of possibilities. The same can be said of my own personal evolution.

I'd always loved pure mathematics—as a high school student on the math team at Stuyvesant High School in New York City, as an undergraduate in Princeton's extraordinary math department, and as a graduate student at Oxford University working on the esoteric field of algebraic combinatorics. However, I knew in my heart that pure mathematics would be too monastic a career for me. I wanted to do something more worldly, but I had no idea what.

While finishing my Ph.D., I somehow wangled a position on the faculty of the Harvard Business School teaching managerial economics (despite having only a passing acquaintance with the field), but soon concluded that economics was not my passion. As I cast about further to find my intellectual passion, my brother Arthur, who was then doing his medical and doctoral work at the University of California, San Francisco, suggested that I might like to learn about the brain. Late in the spring of 1982, he sent me two papers applying mathematics to the cerebellum. Having time on my hands over the summer academic holiday, I set out to try to decipher them. To do so, I found I needed to know some cellular neurobiology. But this depended on cellular biology, which itself required molecular biology, which, in turn, rested on genetics. Being hopelessly naïve, I set out on this regress. During the next three years, I sat in on miscellaneous biology courses around Harvard and MIT and began moonlighting in genetics labs (while still dutifully teaching managerial economics). One afternoon in early 1985, I was introduced to an MIT biologist, David Botstein, who accosted me with questions about analyzing the human genome. I found myself completely captivated, and I've never looked back.

I had stumbled into the most exciting scientific frontier of our age. Within a year, the scientific community began hotly debating the idea of a Human Genome Project—a massive effort to gather and interpret all the information in the human DNA. I found that my combined background in mathematics, business, and biology was an ideal (if completely accidental) fit for the task. I soon moved to the Whitehead Institute and MIT and launched one of the first genome centers in 1990. Since then, I've had the extraordinary pleasure (and occasional frustrations) of working on the greatest collaborative enterprise in the history of biomedicine and being part of one of the greatest intellectual revolutions in science and medicine.

Biology has undergone a nearly complete transformation in its worldview during the past 20 years. In the early 1980s, the focus was on studying each individual component of the cell—a gene, a protein, sometimes just a subdomain of a protein. As with the blind men and the elephant, this provided a terribly limited picture. But it was just impossible to study the whole elephant.

All that has changed. The Human Genome Project has given us the complete DNA code for the human being. While this work has clear utilitarian value, its most important impact has been intellectual. It has made us realize that biology and medicine are *finite* and *tractable*.

In the late 1800s, chemists realized that all matter could be understood in terms of a finite set of about 100 building blocks: the periodic table of the elements. This realization transformed chemistry, setting the stage for practical applications (such as the chemical and pharmaceutical industries) and for a deep understanding of matter (such as atomic theory and quantum mechanics).

Biology is now undergoing a similar transition, and the intellectual impact will be extraordinary. In the coming decades, we'll have a complete accounting of all the genes and proteins; all the regulatory switches that turn them on and off; all the common genetic variation in the human population, which underlies susceptibility to disease; all the pathways that underlie the workings of cells; and all the ways that those pathways can go awry to cause cancer and other diseases. And we'll have chemical tools for modulating all of the genes, proteins, and circuits. It doesn't mean that we will cure all disease, but there'll be tremendous progress. I don't underestimate the magnitude or complexity of the work ahead, but the drive is now inexorable.

The Human Genome Project also taught us about the importance of teamwork. Scientists tend to be individualists, which is crucial in any creative profession. But it has become clear that some of the important goals can be accomplished only if we work together. It's an important cultural shift. I could not imagine a more wonderful time to be in science.

Nicole Le Douarin

Developmental biology. Professor, Collège de France.

As a scientist, you have to be in agreement with what you observe and not with what you think beforehand. I have always been interested in the development of the embryo because it is fascinating to me that a single cell can give rise to such a complex organism as a human being. I like the challenge and the call on the imagination required by research.

I grew up in Britain, where my mother was a schoolteacher and I was one of her pupils in elementary school. I am immensely grateful for the education she gave me. Throughout my childhood and adolescence, her example inspired me and gave me the ambition to learn and to try and understand the complexity of the world.

My scientific activity has been devoted to studying the embryogenesis of the nervous and immune systems. Through my work, I have learned that certain cells which will form the embryo display a very special behavior because they move away from the place in which they originate to contribute to distant tissues. This typical feature was not fully realized before I devised a technique which allows us to follow the migration and fate of embryonic cells throughout development. Applying this cell-marking technique, based on the construction of chimeras (which means associating cells from two different species whose cells can be recognized) in an avian embryo, sheds light on unsuspected events, which are crucial in the fields of cell and developmental biology as well as in evolutionary biology.

A large part of my work concerns the neural crest, a structure which evolved when the vertebrates started to appear during the process of evolution. With my colleagues, we showed that the neural crest plays a major role in the formation of the vertebrate head, a part of the body which evolved considerably during the life history of this group of animals. Some of our findings gave clues about the origins of anomalies of head development in humans. Other studies on the development of immunity, a protective system against infections, involved me in the mechanisms related to tolerance to "self," an important process without which autoimmune diseases develop. The investigations carried out in my group have aimed at understanding the mechanisms of development on a very fundamental level, and much of the knowledge acquired in this way underlies progress in medicine.

Richard Leakey

Paleoanthropology, human origins. Former Director, National Museum of Kenya and Kenya Wildlife Service.

Like many children of famous parents, I decided as a teenager that I definitely did not want to follow in the footsteps of mine. My parents were Louis and Mary Leakey, the eminent paleontologists who discovered many ancient hominid bones in Tanzania's Olduvai Gorge. At age 17, although I had no idea what I might do for a career, I thought I knew what I did not want to do. I had no doubt that I should avoid an academic career and distance myself from my parents and their work on fossils and prehistory, largely because I wanted to be my own man. And so I dropped out of high school and spent the next few years collecting animal skeletons, which I sold to universities and museums around the world, conducting photo safaris, and learning how to fly.

It was flying that indirectly brought me back to paleontology. In 1964, I led an expedition to a fossil site that I had seen from the air. I realized then that I enjoyed looking for fossils. Lacking the scientific education for the task, I went to England for further study and then returned to Kenya to work as a paleontologist. My expeditions have found a number of hominid fossils that contribute significantly to the picture of human evolution. Among them are KNM-ER 1470, a *Homo habilis* skull with an estimated age of 1.9 million years, and KNM-ER 3733, a skull of *Homo erectus* aged 1.7 million years. Paleontology turned out to be the first of several more or less parallel careers. I began an association with the National Museum of Kenya in 1968, when I was 23, and two years later I became the director of the museum. In 1989, the president of Kenya appointed me director of the Kenya Wildlife Service. During my five years in that post, I worked to curb the rampant poaching of elephants in the nation's parks and wildlife reserves. Envisioning tourism as an important contributor to the Kenyan economy, I encouraged community development programs giving people living near the reserves a stake in tourism.

Next was a venture into politics. With others of like mind I founded a reformist political party, Safina, and in 1997 I was elected to an opposition seat in the Kenyan parliament.

Now, in an association with the Great Apes Survival Project established by the United Nations Environment Program, my focus is on an international effort to stem the decline of the great apes—the gorillas, chimpanzees, bonobos, and orangutans. The world must wake up to the fact that the poor nations of Africa and Asia cannot bear alone the financial burden of this effort.

Torsten Wiesel

Visual system. Nobel Prize in Physiology or Medicine, 1981, for discoveries concerning information processing in the visual system.
President Emeritus, Rockefeller University.

I was born in Uppsala, Sweden, in 1924, the youngest of five children. My father was chief psychiatrist and head of Beckomberga Hospital, a mental institution located on the outskirts of Stockholm. Growing up in this environment with daily contact with the patients no doubt influenced my decision to go to medical school and fostered my interest in neuroscience. After medical school and some clinical experience in child psychiatry, I felt a need for more knowledge of brain functions, and from that time on, my focus has been on system neuroscience and brain development.

In 1955, I had the incredible luck to be invited to the United States as a postdoctoral fellow in the laboratory of Stephen Kuffler, then at Johns Hopkins Medical School. Kuffler, who had just published the classical work on the receptive field of cat retinal ganglion cells, soon became my mentor and remained a role model throughout my career. The second stroke of luck came in 1958 when David Hubel joined the laboratory. The two of us then set off on a 20-year exciting journey from the retina through the lateral geniculate body to the primary visual cortex, analyzing the structure and function of this part of the visual system at a single-cell level.

The cat and monkey visual cortex revealed some very intriguing information as we probed single neurons with David's tungsten microelectrode; we discovered that a given cortical cell was, to our amazement, specific in responding only to contours of a specific orientation and, in addition, that the majority of cells were binocular but preferring the left or the right eye in about the same proportions. Furthermore, cells with similar properties were organized in columns, which we called ocular dominance and orientation columns, using a term first formulated by Vernon Mountcastle to describe the organization of cells in the somatosensory cortex.

We conducted experiments to trace the development of such intricate cellular properties and how cells with similar properties came to aggregate into columns. We knew that children born with cataracts never recovered full vision after removal of the cataracts and fully restored optics. Next we wanted to know why the children operated for congenital cataracts had poor vision, since the circuit should have been there at birth. From these experiments, we learned that early in life there is a critical period during which time the neural connections present at birth can be lost or modified by visual deprivation. From our studies, we should not, however, draw the conclusion that the visual system or other parts of the brain cannot make changes in neuronal circuits all throughout life, but that the emphasis should be on the fact that early in life the neuronal pathways are highly sensitive to environmental influences. Nonetheless, the neural bases of memory, learning, and consciousness remain questions to be answered by future generations of neuroscientists.

After 40 years in the lab, I was asked in 1991 to become president of Rockefeller University. Unlike being a working scientist, being university president for seven years provided an opportunity to interact with scientists in many different fields and broadened my scope of the natural sciences. For example, it allowed me to take the initiative to create the Center for Studies in Physics and Biology. To my surprise, I enjoyed the challenges of administration, in particular assisting in the recruitment of new talent to the university. It was a privilege to be at the helm of such a great institution and to be a part of this unique community of scholars and university staff alike.

Since leaving the presidency in 1998, I have devoted myself to programs for younger scientists, in an attempt to offer the same opportunities I had been given over the years. I accepted, in the spring of 2000, the position of secretary-general of the Human Frontiers Science Program Organization, which is headquartered in Strasbourg. The program provides international grants and fellowships, which are highly regarded in the scientific community. The grants are unusual in the sense that their purpose is to provide opportunities for scientists from different disciplines and continents to collaborate. In addition, I continue to be an adviser to different countries, mainly helping to create opportunities for young scientists to carry out independent research, similar to what has become the established path for scientists working in the United States.

Andrei Linde

Cosmology. Professor of Physics, Stanford University.

I was born in 1948, in Moscow, in a family of physicists. But as a kid I did not want to be a physicist; I wanted to be a geologist, discovering treasures of the earth. Everything changed in the last years of middle school. My parents took me in a car trip from Moscow to the Black Sea, and they gave me a book on astrophysics and another one, on the special theory of relativity, to read in the backseat. And when we finally arrived at the Black Sea, all my childish dreams were shattered; I realized that I could not study the earth when there are such wonderful things as red giant stars, white dwarfs, and a mysterious world of elementary particles. I became a physicist.

Physicists of my generation were very lucky to start their active scientific work in the beginning of the 1970s, when the unified theory of weak, strong, and electromagnetic interactions was born. This was a real revolution in the theory of elementary particles, but the results of this revolution did not fit well into the theory of the evolution of the universe. We were forced to think hard, trying to resolve the problems of the standard Big Bang theory. Eventually, we changed it by developing what is now called inflationary cosmology.

This theory, which describes an exponentially fast expansion of the universe at the very early stages of its evolution, gradually became a standard cosmological paradigm. It solved many complicated problems of the usual Big Bang theory. It also made several predictions, which later were confirmed by cosmological observations.

One of the unusual features of our universe is its amazing uniformity on a very large scale. The only explanation of this property is related to inflation. But then we have found that the same mechanism that makes the observable part of the universe so uniform simultaneously produces small inhomogeneities, which grow up and give rise to galaxies. Moreover, on a much larger scale, beyond what we can observe now, the same mechanism leads to the formation of new parts of the universe. We have found that instead of looking like a spherically symmetric ball, the universe may look like an eternally growing tree consisting of an infinitely large number of balls producing new balls, with different laws of physics in each of them. Even though each of these balls may collapse and disappear, the universe as a whole, according to this theory, is immortal. I called this theory "eternal inflation."

Some people believe that cosmology is the ultimate science: it describes everything that can possibly exist. Being a cosmologist, I share this point of view, but only up to a certain extent. I believe that the ultimate science should describe life rather than the place where life can exist. But by looking at our cosmic home, we can say a lot about its inhabitants. And, vice versa, by studying ourselves, we can say a lot about the universe where we live now.

Susan Lindquist

Stress response, protein folding. Professor of Biology, Massachusetts Institute of Technology.
Director, Whitehead Institute for Biomedical Research.

My fifth-grade teacher walked into the classroom one day and said, "Close your books. Today we're going to try to answer the question 'What is life?'" We started listing things on the board. Something is alive if it moves, eats food, uses oxygen. She said, "What about a car? Doesn't it do those things? Is a car alive?" We spent maybe an hour on it. But I continued to think about that question as the years went by. And I continue to think it is the most compelling intellectual question there is. I've been asking that question—on a variety of different levels and from many different angles—throughout my career.

I came to a career in science, however, through a series of happy accidents. When I was a little girl, I used to imagine ways in which I could help mankind. I read about Jane Addams and Hull House, and I wanted to be a social worker. Then I wanted to be a nurse. Then—a bold thought in my household—I wanted to be a doctor. I am a second-generation American; my mother is Italian, my dad Swedish. They were very warm and loving people, but in the 1950s and 1960s the concept of a girl's wanting to become a doctor seemed ridiculous. So even as I started having dreams of having a career, I never really took myself seriously. I did take several science courses in college, but I have to admit that I spent an awful lot of my time making friends, dating boys, and having an absolutely great time. Then one of my teachers encouraged me to think about a career in biological research, and I thought, "Wow! You really think I could do that?" It's been a progressive series of awakenings ever since, and not just to the world of science but to my internal self in terms of what I'm capable of doing. We've tackled a variety of different types of problems in the work we do in the lab, and that constant growth has been really exciting.

Our work has always been centered on trying to understand the very fundamental, basic processes of biology. My conviction is that these understandings will wind up getting leveraged in many beneficial ways that can't be predicted. Paradoxically, as the biological sciences have become so much more sophisticated and technology driven, we also come to appreciate some simple things: that the basic life processes of all organisms are very closely related. That means that we can move from one system to another and directly study some very complex medically relevant processes in simple organisms. In fact, we are now studying some terribly complex problems—neurodegenerative diseases that are due to misfolded proteins—in yeast cells! The reason we can do so is that the problem of protein folding and misfolding is as ancient as life itself.

I came into science for the sheer intellectual joy of it. I am simply awed to find myself a participant in one of the most extraordinary events in human history. We're actually figuring out the answer to the question "What is life?" And the more we learn, the more beautiful, and crazy, and amazing, and fascinating, and wild, and wonderful it seems. The downside is that everything is moving so fast. Keeping up with it is exhausting. The upside is the sheer wonder of it all. It's one constant surprise after another. Science is not an easy profession, but now that I've tasted it, it is the only one I can imagine following.

So I find myself working at the remarkable intersection of the two things that meant a lot to me as a young girl. This extraordinarily exciting intellectual quest is also of fundamental importance to humanity. We are facing some extraordinary problems. We already have many horrific diseases, and new ones are emerging. The threat of bioterrorism is very real. We are polluting our environment and overpopulating the planet. People are starving to death, and other species are disappearing from the earth at an extraordinary rate. These biological problems—and our abilities to solve them—will shape our destinies. For the sake of my children, whom I adore, and their children, whom I can as yet only contemplate adoring, I hope we succeed.

Roderick MacKinnon

Bioelectricity. Nobel Prize in Chemistry, 2003, for structural and mechanistic studies of ion channels.
Professor of Biophysics, Rockefeller University, and Howard Hughes Medical Institute.

I think I had the makings of a scientist from a very young age, in that I was always very curious and always wanted to know how things work. When I was in elementary school, my mother signed me up for a summer enrichment course in science. They let me take a tiny microscope home, and I spent hours looking at things in pond water, blades of grass, and leaves. I was fascinated by how things work in the world around me, and I suppose that was a sign that I had the makings of being a scientist. In college, I was a biochemistry major because mathematics and science came naturally to me, but I didn't think a lot about what I wanted to do. A lot of my friends were going to medical school, and I did, too. I was naïve in going. I thought medicine was a kind of science, but it's different. It's a professional school where you learn a lot of information, and I discovered I was better at solving problems given a few rules.

I enjoyed the clinical years of medical school and went on to do a residency in medicine. During the residency, I realized it wasn't for me. I could do it, but I wasn't thrilled. It didn't make me feel the same way I felt if I thought about a little problem and solved it. At the end of residency, I left medicine and went to do postdoctoral work at age 30. I had a lot of catching up to do, so I studied intensely at night and when I wasn't doing experiments. I developed a sense of ease about picking up new subjects and learning them. That's paid off well in my life as a scientist because I'm not afraid to study something new. You get books. You learn. You talk to people. You pick up techniques, and you do it!

My postdoctoral work was on proteins called ion channels. They are the proteins that make the electrical signals in us, in living systems. They are the basic hardware for the electrical impulses that let us move and think. In my reading, I saw that ion channels are very simple devices that do something quite simple: they conduct ions. The electricity made by cells is made by ions, which are charged atoms moving across the cell membrane. The atomic details were not known, but the basic physical principles seemed likely to follow simple rules. It was the interface of biology and physics, and that appealed to me. What people didn't know is what atomic structures make the molecules that make this happen. That is what fascinated me and led me to where I am today. We have been able to capture the first pictures at atomic level and tell you quite precisely how the charged atoms (the ions) move across the cell membrane. We have been able to show the molecules that make electricity in living cells. So I'm a photographer, too! My subject's just a bit smaller. We have to use a computer lens, and we have to deduce what the lens should do to the rays to turn them into an understandable image.

Why would anyone care about such a pure intellectual pursuit? Because it's fascinating and it's the world around us. I am an explorer. If I think about wiggling my finger, somehow a message is conceived in my brain. The message gets down to the muscles in my finger to make it wiggle. How? By an electrical impulse carried along the membrane of nerve cells and that membrane has ion channels in it that allow it to make the electrical impulse. Ion channels are related to movement, thoughts, the beating of our heart. That means abnormalities like arrhythmias and epilepsy have to do with abnormal ion channel function in some cases. There are already some drugs that work on different ion channels—potassium, sodium, and calcium channels. I always want to emphasize that my work is foundational. I want to understand what these ion channels look like and how they work. It is still several steps away from application, and it will take a lot of work of others to bring some of what my lab does to application. It would have to do with many nerve diseases. There are many processes controlled by the electrical impulses: blood pressure, the airway tone, hypertension, and asthma. If my work has led to an advance, it has been on a very basic level of understanding the way cells produce electrical signals. What's beautiful is that if you look at the basic hardware (i.e., the potassium channel, which conducts potassium across the cell membrane), it's very similar in people, animals, plants, and bacteria. These molecules are very basic to life. I enjoy my work.

Marek Zvelebil

Human prehistory. Professor of Archaeology, University of Sheffield, UK.

I discovered the past and the puzzle and mystery it captured quite suddenly one day when I was walking through the forests of my native Bohemia with my father. We stumbled across ruined foundations of a medieval fort or a castle, one that was not marked on the maps. How did it come to be there, what did its ruins mean, what events came to pass there, and why has it been so comprehensively forgotten?

At that time, the country was a communist state, and in the socialist Czechoslovakia of the 1950s and 1960s schoolchildren were sent off in the summer to "young pioneer" camps for political indoctrination, to train in the "building of socialism." The day in such camps usually began with patriotic and revolutionary songs, marches, and trooping in front of the camp leader, who would shout, "For the defense and development of the fatherland, be ready!"—to which one had to yell back, "Always ready!"

All this was extremely boring and embarrassing. At the age of 14 I discovered that if I volunteered to work on archaeological excavations, I could do this instead of being a pioneer. So I spent several happy summers digging for archaeology, in a much freer environment, which was intellectually stimulating too. Practical work in the field and intellectually provoking research form the basis of archaeological investigations: this is something fairly distinctive to archaeology as a discipline, and it is one of the main reasons why I remained captivated by archaeology for the rest of my life.

Because it studies material remains of past human communities, of people and individuals, archaeology is a spatial and humanist discipline. People do not live their lives in isolation, they live in geographical space, and their lives are conditioned by the experience of previous generations. Archaeological investigations then must be informed by geography, history, and anthropology and anchored in space and time. My own methodological approach tries to follow these convictions: in the field, I practice landscape archaeology, which combines excavations of sites with broader landscape research in order to understand people in their environment. I then combine archaeological information forthcoming from the analysis of artifacts, structural remains, and settlement patterns with findings from biological and ecological sciences. Here the unique nature of human species comes clearly through: as biological organisms, we share our adaptive responses with other species and react physiologically to our changing environment; as human users of material culture, we have developed an alternative route to survival: an incredibly varied range of cultural responses, adjustments, and innovations. These in turn have changed our species biologically and intellectually. And all this has been going on within the constraints and possibilities created by the historical experience of previous generations over thousands of years!

My research has focused on one major historical problem: the invention and development of agriculture. The practice of farming changed our lives dramatically since its invention about 10,000 years ago. In many regions of the world, it allowed human populations to grow dramatically. Higher population densities, changes in dietary patterns, and close contact with animals brought new diseases and epidemics to humanity. Larger settlements and closer cohabitation required new social arrangements and ideologies—mostly toward social ranking, social elites, and appropriation of resources. Farming also required a great deal more work than did the hunting and gathering that people had been practicing for over a million years prior to farming. For less than 1 percent of our hominid history have we lived as farmers, yet today less than 1 percent of people are still engaged in hunting and gathering. Such was the success of the transition to agriculture. So why did people adopt farming?

Different scholars hold different views, and there may not have been a single, universal cause for the adoption of agriculture—after all, farming covers a very broad range of practices. My research leads to the conclusion that in temperate, nontropical regions, at least, shifts to greater sedentism and to social ranking encouraged the transition to agriculture (by this I mean mainly cereal-based farming and stock keeping). Both of these shifts began among the hunter-gatherer communities, creating situations where transition to farming would prosper.

My understanding of the agricultural transition is not shared by everyone. But few scholars would disagree with the view that the invention and practice of agriculture profoundly changed our human existence. This is why Gordon Childe called it the Neolithic Revolution. And yet patterns of human behavior that encouraged and promoted the development of farming were already present among the more settled hunting-gathering communities prior to its development. They included increase in reproduction and population growth, greater control over natural resources, and increase in social competition and ranking. The adoption of farming accelerated and transformed these existing patterns of behavior to a new, more complex level, which we commonly associate with civilization.

Geoffrey Marcy

Planetary systems. Professor of Astronomy, University of California, Berkeley.

One day when I was 14 years old, my parents bought me a used telescope. That evening, I looked at the planet Saturn and could not believe my eyes. There crystal clear was the ring encircling Saturn, a billion miles from Earth. I was hooked.

During the next few years, I looked at Jupiter and its four Galilean moons, the Orion Nebula, where stars and planets are being born, and the Andromeda galaxy, located a million light-years away. I was stunned at the untouched beauty and the vast distances of the universe, making me feel gloriously insignificant. I sensed that we humans, living our brief, frenetic lives, might find some peace and wisdom knowing that our origin and destiny lay in the stars.

I learned of another type of beauty in high school and at the university. There I learned that a succinct set of physical laws allows us to anticipate exquisitely the behavior of nature. Atoms, guitars, and spiral galaxies are governed by the same fundamental laws of nature, and their mathematical expression can be written on a postcard. Lightning strikes and brain activity are next of kin. Compass needles and our unconscious feelings stem from the same physical effects. How connected everything seems!

I studied math, physics, chemistry, and astronomy at UCLA, working on homework till ten each night, after which I learned even more from late-hour talks with my friends in the dorms. But my favorite course at UCLA was cultural anthropology, where my professor explained that all civilizations adopt their own cherished cultural values and religions just as we do today. Humans across the globe and throughout time have struggled similarly with rites of passage, mating rituals, and searches for meaning. This commonality of human cultures seems reminiscent of the universality of physical laws. Looking both outward and inward might reveal deeper understanding. I wanted this.

In graduate school, I learned to do research, but was crippled by self-doubts. I felt I wasn't smart enough to be a scientist. Luckily, some insightful people told me to care less about the opinions of others and just go with my inner desires. One morning, I got into the shower and just stood there under the water wondering what astronomy research I should do. Suddenly the idea hit me that I wanted to know if Earth was unique or if instead it was a common type of planet in the universe. Maybe Earth and its cultures were a part of the universal laws.

I decided to dedicate my life, starting in 1984, to finding other planetary systems around other stars. During the next 20 years, my dear collaborator Paul Butler and I developed a novel technique for discovering new worlds around nearby stars.

We've now found 87 planets and are still counting. Some systems are completely different from our solar system, with its nine planets in circular orbits. Some of the new planets move in elongated orbits, egg-shaped orbits that look nothing like our solar system. Some planets orbit so close to their host star that they are blow-torched to thousands of degrees. And some planetary systems resonate like chimes, with each planet orbiting with a frequency coupled to its neighboring planets.

We have found planets the size of Jupiter and Saturn, but we can't yet detect Earths. These are simply too small to cause the host star to wobble, our key to planet detection. But we see more small planets than large ones, suggesting that nature makes even more of the smaller, Earth-like planets. Our Milky Way galaxy probably harbors over 10 billion Earths. Some are no doubt too frigid for life, and some too hot. But some of the billions of Earths must be bathed at lukewarm temperatures and must have atmospheres filled with the same atoms and lightning strikes as our Earth.

There, the universal physical laws must surely produce replicating molecules similar to DNA, making primitive life highly probable. The competitive nature of evolution must breed ever more complex creatures. Who can doubt that some of those extraterrestrials struggle with their own rites of passage, mating rituals, and the meaning of their existence? Wouldn't it be glorious to meet our cosmic kin and compare notes?

Lynn Margulis

Evolution. Distinguished University Professor, Department of Geosciences, University of Massachusetts, Amherst.

In a school essay I wrote, "I was born on this University of Chicago campus, I attended The Laboratory School a few blocks away and now I'm still around as a student in The College so I expect to die here too," which inspired my teacher to mark in the margin, "Don't be so macabre!" My academic experience at that premier institution has never been rivaled, but the calm green campus and the intellectual passion of the university is exceedingly anomalous in the huge city that surrounds it. Indeed, I spent the first 19 years of my life in an attempt to escape the filth, racism, and pretension of the urban-blighted South Side. Perhaps I love nature so passionately because on my home range, between 69th and 71st streets, South Shore Drive and Jeffrey Avenue, there was such a paucity of it. My father's three-flat apartment building was separated from the Great Lake by an exclusive golf club. The conspicuous sign on the iron gate entrance actually read: "No Negroes or Jews allowed." By only a two-block detour, I could reach the shore. Often frustrated but never lonely, I spent hours climbing on its rocky breakwater. Other options were scrambling within the high maze of connected garage rooftops that formed a secret second city or exploring the tree greenery of Jackson Park. Besides the reading of books, everything I liked then, and still do, tended to be more akin to the favorite activities of a ten-year-old boy than to those of the grandmother I am.

I did not participate in science until I was forced, at the College of the University of Chicago at age 15, to study Natural Sciences 1 (physics and chemistry). The following year in Natural Science 2, we asked, "What is inherited?" How does a sperm-penetrated egg become a child? What is the material transmitted through the sperm and egg? It was the enthusiastic inquiry style of Natural Science 2 that inspired me. We pondered the chemical basis of heredity by reading choice professional scientists and learned that one could design and carry out experiments to learn at least one small thing. I could immerse myself in observations and experiments of science's illustrious predecessors. What a revelation: to question authority became necessary! Science encouraged us to demand evidence; it bade us avoid pomposity and knee-jerk truths, to query the accepted truths of politicians and religious leaders. We were introduced to an international community of scholars mindful of evidence. Their everyday activities were built on values of curiosity, honesty, and the need for detailed description. This was, to me, the realization that I enjoyed allies. Always skeptical, even disdainful, I had doubted what talkative folks told me (e.g., white people were superior, God made flowers, swimming nude was sinful, virginity was a prerequisite to a stable marriage). But not until I began to *do* science did I discover that I, too, could participate in a fascinating exploration of the world. From those college years until now, I have never been more content than when, with knowledgeable and committed students and colleagues, I study nature from within by means of science.

We study the evolution of cells, the units of which all life-forms are composed. Only two kinds exist: cells are either bacterial (small and lack nuclei), or they contain nuclei and are larger. All plants, animals, fungi, and their ancestors (protoctists, a diverse group of organisms, some tiny and some, like kelp, huge and multicelled) are composed of cells that invariably contain membrane-bounded nuclei. How did plant and animal cell ancestors evolve from protoctists? How did protoctists evolve from bacteria?

With colleagues, by detailed study of marine muds, pond water, and intestinal microbes in wood-eating termites and roaches, I've studied these problems for over 40 years. As eclectic evolutionists, we apply techniques selected from biochemistry, cell biology, computer analysis, genetics, microbiology, and paleontology. We study bacteria and protoctists from insect guts, from brightly colored muds from salt flats, and as remains trapped as ancient fossils in rocks.

I suspect close collaborators (Michael Dolan, Richard Guerrero, John Hall, and Dennis Searcy), and I have figured out, at least in outline, how the first nucleated cell originated. Probably on an ancient Earth prior to 1,200 million years ago (during the Proterozoic eon) when the air lacked much oxygen gas. The oceans then, like today's Black Sea, were full of sulfur in oxygen-poor waters. The earliest nucleated cells did not breathe oxygen but thrived in sulfide-rich water we humans and other mammals would consider fetid.

Reconstruction of evolutionary history uses a vast scientific literature; clues from the living are taken to be representational of past events. Teeming bacteria of different kinds fused their bodies. The nucleated cell is far more a permanently merged set of diverse bacteria than a bacterium that complexified by random mutation and grew larger. In our videos, microbial symbionts fuse their bodies; they form more complex beings. Dorion Sagan and I detail these ideas in our 2002 book, *Acquiring Genomes: A Theory of the Origins of Species*. Evidence for our scenario abounds, and the search for even more new evidence continues to delight us!

Ernst Mayr

Evolutionary biology. Alexander Agassiz Professor of Zoology, Emeritus, Harvard University.

I was a very lucky youngster. Both of my parents were naturalists and taught me and my two brothers what the local birds were, as well as their songs, the spring flowers, the fossils in limestone quarries near our hometown, and whatever interesting diversity nature produced. I was a voracious reader of the travel books of the great explorers from Humboldt, Alfred Wallace, and Bates to the explorers of the Arctic and Antarctic, as well as those of inner Asia. I was dreaming all the time of undertaking also such an expedition and making great discoveries.

Totally unexpectedly, through the efforts of the great Berlin ornithologist Erwin Stresemann, I was sent out to the tropics by Lord Rothschild and spent two glorious but also highly dangerous and hardworking years in New Guinea and the Solomon Islands. What I learned on these expeditions made me a specialist in bird classification and in such branches of biology as systematics and evolution.

The period of my early activity in these fields coincided with the great controversies of the 1920s to 1940s about evolution—that is, about the origin of the astounding diversity of nature, from minute unicellular creatures to giant trees, from birds of paradise and hummingbirds to orchids and cereals.

Evolution in due time became my field of specialization. How did the species originate that were the components of this diversity? I found that there are two major fields of evolutionary biology. One is the study of local populations, their composition of genes and the change of these gene pools from generation to generation. I learned that there was a constant production of new genes (gene mutation) and the sorting by natural selection of the variation that is produced in every generation. This had been clarified in recent years by a number of great geneticists, but they had hardly touched another, equally important field, the origin and maintenance of the enormous biodiversity. How are new species produced? How should one define species? The geneticists did not know how to deal with this problem. It was up to the naturalists, like myself, to do so. I helped to develop a new species concept, the so-called biological species concept, based on the criterion of the successful interbreeding of populations. Having this new kind of species concept made it possible to show how isolated populations can acquire enough genetic differences to become reproductively isolated from other populations, that is, new species.

In time, I showed that individuals and populations in the living world have characteristics, like genetic programs and biopopulations, that do not exist in the world of physics. This necessitates the development of a new philosophy of biology, which is in principle different from the philosophy of physics. It has been my endeavor in recent years to contribute to the development of such a philosophy of biology.

Vera Rubin

Dynamics of spiral galaxies. Astronomer, Carnegie Institution of Washington.

I am an astronomer because the beauty and the mystery of the night sky captivated me as a youngster, and I could not imagine living on Earth without trying to understand what I was seeing. Even now, the sight of the bright stars against the dark sky from a remote mountaintop observatory is the most remarkable view I know. Just as I could not be happy living on Earth not knowing that there were continents and oceans and what a map of Earth looks like, I was not content until I knew the connections between galaxies and stars and planets, and what the map of our corner of the universe looks like.

So I do science because I fell in love with a way of life that would permit me to be a perpetual student, to learn what is known about the universe, and perhaps to learn some things never before known. It was the beauty, the unlimited scope, and the cumulative structure of science that made astronomy my career choice.

Combining a life in science with an active family has been possible and fun because of the support and encouragement of my mathematician/biologist husband, Bob, and because of the opportunities offered me by the Carnegie Institution of Washington. Allan, the youngest, recently remembered that as a young child, he would occasionally ask where Mother was. And the reply "She's observing" reassured him, for everyone seemed content even though Allan did not know what "observing" meant.

Summers combined family and work at interesting sites: Los Alamos, Boulder, Brookhaven, Lowell Observatory, Kitt Peak National Observatory, McDonald Observatory. Bob and I did not know that picking up rocks would turn Dave and Allan into geologists; or that taking a course that her mother was volunteer teaching at the local high school would turn Judy into an astronomer; or that the math puzzles in the car, coupled with the many math books in the house and discussions with Bob, would make Karl a mathematician.

My career has been spent learning how stars orbit the centers of their galaxies, and how galaxies move in the universe. These subjects were chosen partly because I was curious about the outer regions of galaxies, partly because these areas were not being actively studied by others. This would permit me to work at my own pace, which seemed important. From the large orbital velocities of stars and gas that I detected, we learned that most of the matter in a galaxy is dark. This "dark matter" is not radiating at any wavelength, but its gravitational effect on the bright stars causes the stars to move at unexpectedly high velocities. In the 1930s, Fritz Zwicky had concluded that dark matter existed in clusters of galaxies, but this result was not then widely embraced. With the new evidence from high rotation velocities in galaxies, arguments for the existence of dark matter were persuasive.

Astronomers now know that at least 90 percent of the matter in the universe is dark. It envelopes galaxies; it extends far beyond their optical image. We do not yet know what it is composed of. It is curious that I have found the universe to be even more mysterious than it seemed when I puzzled about it as a child. It is also funny that an optical astronomer should end up studying something that she cannot see.

Philip Morrison

Physics, search for extraterrestrial intelligence. Institute Professor, Emeritus, Massachusetts Institute of Technology.

About the oldest memory I can date is a trip to New York City from our home in Pittsburgh. What I best recall is my father rolling me around to look up at the museum ceiling where hung a dusty full-scale model of a giant squid. It set me up for a lifelong interest in squids—not much to do with what I actually worked at. I was four or five years old on a medical trip, a polio wheelchair patient at three, unable to walk until my seventh year.

My dear mother and father took warm care of their limping son, who was a hungry reader and a sharp talker from early on. At five years, Dad brought me home a crystal set, sold to build an audience for the pioneer broadcast station KDKA, only a mile or so from where we lived. Radio caught me even more than squids! First I made toy radios of clip art and scrap in boxes, and the block kids and I made whole dramas out of them in the scrubby bush nearby. By eight or ten, I was a builder of working shortwave receivers, by twelve I was a licensed amateur. How happy I was when my five-watt transmitter drew an answer from an Irish station!

When I entered Carnegie Tech—a streetcar day student on full scholarship in Depression years—I wanted to be a radio engineer. But I met the physics faculty, and they were so full of questions about how things really worked that I became a physics major instead, and have never turned back.

That gifted redhead Galileo Galilei of Florence recognized the coming of a new astronomy once he heard of the Dutch optical tube, and soon developed his own to better the foreign import. He was able to map at the eyepiece a dark region quite free of more than a very few visible stars to reveal unmistakably scores of them. The moons of Jupiter had been feebly glimpsed once or twice by keen-eyed watchers before him. But their endless complex dance, their shadows, their eclipses, and their obedience to the bright planet's wide orbit disclose a whole new astronomy, hidden only by faintness, but there to be seen with care, attention to quantity, and a few subtle glass eye-aids. It would entice pope and prince to learn.

I say in all earnestness that the astronomy and cosmogony we test today, using an arsenal of new channels, and a powerful theoretical coherence to extend our subtle propositions far beyond the fall of unlike weights, is a revisit that will transcend the Tuscan past. For within the last 10 years we have been able to set beyond doubt big planets in orbits that resemble our solar system though held to an alternate sun nearby, but 100 and more of them. An extrapolation I can defend—not clearly certain, but by no means unlikely—suggests that the Milky Way galaxy we inhabit contains in total a million times more exoplanets. Those other worlds, other suns are in multiplicity that fits the long-tested account that gave us something like a tenth of a trillion stars in our own cosmic patch.

What will astronomy be like when such objects are surveyed and assessed? Surely half of our successors will study these exoplanets. What a treasure has turned from fiction to fact within one tenure track! Even if they are no more than the 100 now seen within 100 light-years of home, they will draw effort on paths all too easy to propose.

For those who seek the speculative, in really ancient times and at truly cosmogonic distances, the same novelties are here, or at least shelved among newest journals, or maybe only some websites to appear tomorrow. The size and age of the cosmos is open; our local pool is fresh, maybe only 10 or 20 billion years old, though its extent can be extended to any finite limits, since speeds of motion no longer remain constrained by light speed and gravity, but hold strong signs of antigravity without plausible limit at early times in the very clear and simple evidence of widespread uniformity of the large-scale world. Here the facts are still less sure, but this is by no means beyond the evidence of recent work.

So I can cite the tireless Faraday, who taught without advanced mathematics that nothing is too wonderful to be true, provided it remains within the physical bounds we are still seeking to establish.

A space probe called Kepler is now under construction; it is to test our exoplanet census by 2006 or 2007 much better than we can do from the ground.

Watch all the spaces you can!

Matthew Meselson

Molecular genetics and evolution; control of biological weapons. Thomas Dudley Cabot Professor of the Natural Sciences, Harvard University. Director, Harvard Sussex Program on Chemical and Biological Weapons Limitation.

I attended public schools in Los Angeles and, at age 16, entered the University of Chicago, expecting to study chemistry and physics. But Chicago had eliminated undergraduate programs in specialized subjects. All undergraduates had to follow essentially the same curriculum: classical readings in history, philosophy, literature, and the physical and social sciences, plus mathematics through calculus and a modern European language. I was fortunate in this, because the modern American college degree in science can leave one with only a spotty and idiosyncratic knowledge of anything else.

Starting in high school, I wanted to apply the concepts and methods of the physical sciences to the understanding of life processes. That led me to become a graduate student of Linus Pauling at the California Institute of Technology, where my dissertation research included the invention of a method for analyzing the density of giant molecules. With Franklin Stahl, I applied the method to show that DNA replicated as Watson and Crick predicted—by the two strands separating, with each strand acting as a template for the formation of a new partner strand. Subsequently, I employed the new method to show that genetic recombination takes place by the breakage and joining of DNA molecules and, with Sydney Brenner and François Jacob, to demonstrate the existence of messenger RNA.

In 1961, I moved to Harvard. There, working with graduate student and postdoctoral colleagues, I isolated and characterized a number of DNA restriction and modification enzymes and demonstrated methyl-directed DNA mismatch repair, a mechanism by which mistakes in DNA molecules are corrected. I became interested in the problem of why nearly all animals and plants reproduce sexually, a fundamental unsolved problem in biology. We are taking an experimental approach, trying to find out what has allowed an apparently all-female class of diminutive aquatic invertebrates, the bdelloid rotifers, to evolve successively for tens of millions of years without sexual reproduction.

A different part of my life, starting with a period of government service in 1963, has been work on the control and elimination of biological weapons. Almost every major technology has been intensively exploited not only for peaceful purposes but also for hostile ones. It will take intelligent and far-seeing policies to prevent this from happening to biotechnology.

Leslie Orgel

Evolution, origin of life. Research Professor, Salk Institute for Biological Studies.

I like puzzles. As a teenager I was puzzled by the color of the chemicals in my chemistry set. This led to a first career studying transition metal ions, which are the colored components of a vast array of minerals and solutions. Later I moved on to a much more difficult puzzle: the origin of life on Earth.

The earliest living organisms had to possess genetic information—heritable information for functioning and reproducing. Modern organisms carry their genetic information in nucleic acids, RNA and DNA. So one can infer that the earliest simple organisms stored genetic information in nucleic acids that specified the composition of all needed proteins. Hence the central problem of origin-of-life research can be refined to ask, "By what means did this interdependent system of nucleic acids and proteins come into being?"

Anyone trying to solve this puzzle immediately encounters a paradox. Nowadays nucleic acids are synthesized only with the help of proteins, and proteins are synthesized only if their corresponding nucleotide sequence is present. It is extremely improbable that proteins and nucleic acids, both of which are structurally complex, arose spontaneously in the same place at the same time. Yet it also seems impossible to have one without the other.

In the late 1960s, Carl R. Woese, Francis Crick, and I independently suggested a way out of this difficulty. We proposed that RNA might have come first and established what is now called the RNA world—a world in which RNA catalyzed all the reactions necessary for a simple organism. We also posited that RNA could subsequently have developed the ability to link amino acids together into proteins. This scenario could have occurred, we noted, if prebiotic RNA had two properties not evident today: a capacity to replicate without the help of proteins and an ability to catalyze every step of protein synthesis.

There were a few reasons why we favored RNA over DNA as the originator of the genetic system, even though DNA is now the main repository of hereditary information. One was that the ribonucleotides in RNA are more easily synthesized than are the deoxyribonucleotides in DNA. Moreover, it was easy to envision ways in which DNA could evolve from RNA and then, being more stable, take over RNA's role as the guardian of heredity.

Over the last 20 years a great deal of evidence has lent credence to the idea that the hypothetical RNA world did exist and lead to the advent of life based on DNA, RNA, and proteins. But the precise events giving rise to the RNA world remain unclear. We are still working on that puzzle.

Christiane Nüsslein-Volhard

Biology. Nobel Prize in Physiology or Medicine, 1995, for discoveries concerning the genetic control of early embryonic development.
Director, Max Planck Institute for Developmental Biology, Tübingen, Germany.

I have the fondest childhood memories of being excited by looking at plants and animals: watching and observing and finding out. I have done that all my life. My family was more interested in the arts and music, and I, too, like that very much. But, from very early on, I think I wanted to be a natural scientist.

Biology is the study of how you get from a very simple structure to a complicated structure. This is the essential quality of life. Self-replication is of first importance. Animals and plants—all living organisms—have the capacity to duplicate themselves, and this is something really special. I became interested in genetics, which means using the genes to find the factors responsible for such things. You begin with the assumption that there must be some things in the egg that organize it or tell the cells what they should form, and these factors are probably encoded by genes. When you find the gene that is encoding the factor, you also have access to the factor. Other people tried to find the factors directly; but that didn't work. It's too difficult. I knew bacterial genetics had been very successful, but bacteria do not develop. They are simple cells, which multiply by just splitting. Animals do it differently. They have eggs that develop into a complex organism. I looked for an animal that could do that and chose the fruit fly.

I looked first at the embryos and learned how they develop. You watch and see, observe it, and then describe it. This had been done before, but I developed some new methods of how to look more closely. In the case of the fly, the eggs develop into a little maggot, which is much simpler than a fly. It's just a wormlike structure with segments. In this work I joined forces with Eric Wieschaus. We had questions like "Why is the head always developing at the tip of the egg? And how is the egg divided up into segments?" We looked for genetic changes that would lead to an alteration of the organization of the larva. We had larvae which didn't have a head, for example, or they had fewer segments. These mutations were caused by particular single genes. There were not many of them. We found a hundred or so, and we could describe what they were doing in the animal. This was the work that won us the Nobel Prize.

It turned out—and this is the big surprise of developmental genetics—that you could compare DNA from different organisms. It was discovered that the genes we found in the fly also have very similar forms in humans, mice, and fish. And so, what we first started as a very interesting scientific story, which had only scientific interests, turned out to have a big bearing on general development processes.

Wolfgang Panofsky

Physics, nuclear research. Professor and Director Emeritus, Stanford Linear Accelerator.

I was brought up in Germany; both of my parents were art historians. My father used to designate his two more technically inclined sons as "die Klempner" (the plumbers). Then came Hitler, and my parents left for the United States. I entered Princeton at the age of 15, studying mainly science and engineering, though I had to give a Latin speech at graduation. I moved to California, where I did my Ph.D. work at the California Institute of Technology, in Pasadena. My thesis was entitled "Precision Measurement of Natural Constants using Xrays." Then the war started, and I divided my work among teaching, continued physics research, and war work.

As it happened, Professor Louis Alvarez was asked by Robert Oppenheimer to design devices to measure the explosive power of the nuclear weapons then under development. Alvarez had read a report that we had written at Caltech measuring shock waves from supersonic bullets, and he decided that these devices could be easily adapted to the job he had been given by Oppenheimer. Accordingly, I commuted between Los Alamos and Pasadena, adapting the devices used for shock wave measurement to measure the power of the nuclear explosions. I was on a B-29 only 10,000 feet above the first nuclear explosion at Alamogordo, New Mexico.

After the war, Alvarez persuaded me to go back to Berkeley with him, at the University of California Radiation Laboratory, in 1945. I became his right-hand man constructing the 32 million electron volt (MEV) Proton Linear Accelerator. At the same time, I became very much concerned about the impact that the advent of nuclear weapons had made on the future of civilization, and I spent some time explaining the new source of energy to various lay groups.

I decided to leave Berkeley in 1951, in part because I was upset by the conflict then brewing as a result of the loyalty oath which the regents had imposed on the staff of the University of California. After some hesitation, I moved across the Bay to Stanford. I took leadership of the High Energy Physics Laboratory at Stanford, and a very productive period ensued. Bob Hofstadter did his famous work on the structure of the neutron and proton using electron scattering, and I worked on numerous basic experiments with several graduate students in particle physics. Based on this success, we generated a proposal in 1957 to build the "Monster," now known as the Stanford Linear Accelerator, the two-mile electron machine, which came into operation in 1966 and has been running ever since.

I became the director of SLAC, as the new laboratory was called, once it was approved in 1961, and continued in that function until my retirement in 1984. Through all this time, I continued to divide my time between directing that new exciting laboratory, participating initially in research work there combined with teaching, together with increased involvement in national policy—in particular, arms control. In that latter realm, I became a negotiator in 1959 on behalf of the U.S. government on the delegation in Geneva on how to verify a future treaty on the cessation of nuclear weapons tests, to include such tests not only in the atmosphere but also underground and in outer space.

SLAC requires collaboration among experimentalists, theoretical physicists, administrators, highly skilled engineers, technicians, and nontechnical workers. I was convinced that a laboratory like this could thrive only by a delicate balance between decentralized initiative exercised by many people with good ideas and the direction of the laboratory to make choices among all those ideas, and seeing to it that all the work progressed in the most constructive way. While all disciplines participated in an indispensable way, it became clear that advances in technology were pacing what could actually be accomplished. This continued to be the case after my retirement from the laboratory, and is still the case throughout most of the science.

Advances in technology also have a side effect threatening all humanity. These advances have made it possible that more and more destructive power can now be concentrated in fewer and fewer hands. Nuclear energy has made it possible to amplify that power by a factor well over a million carried by munitions of a given size and weight. All new developments and technologies have been used for both peaceful and military applications and have spread across the globe. This new threat must be managed and controlled. Since my retirement, I have spent the preponderance of my time in various activities related to arms control, the global effort to bring these new technologies under control. It is a challenge for the future to assure that these developments are managed cooperatively worldwide for the benefit of all.

C. R. Rao

Statistical research. Professor of Statistics, Emeritus, and Director of the Center for Multivariate Analysis, Pennsylvania State University.

I come from a family of six brothers and four sisters. All of them have had good academic careers, most probably thanks to good genes we have inherited from our parents and to good upbringing. According to statistics, the second born has a lower IQ than the first born, the third a lower IQ than the second, and so on. I am the eighth child! However, from my early years, I showed some interest in mathematics. At the age of six, I knew by heart multiplication tables up to 20 by 20, and at eleven I could do complex arithmetical problems mentally. My father thought that these were promising signs to become a good mathematician and encouraged me to study mathematics. I chose mathematics as my major subject when I entered college and graduated with a master's degree in mathematics. At that time, the Second World War was on, and it was difficult for a mathematician to get a job. After a few months of waiting and frustration, I found an opening for a mathematician in the Army Survey Unit and went to Calcutta for an interview. As luck would have it, I did not get the job, but discovered the Indian Statistical Institute (ISI), founded by a professor of physics, P. C. Mahalanobis. I was told that those who completed the one-year course in statistics there got good jobs immediately. This seemed to be attractive, and I took admission to the course, which started my association with the ISI that lasted for about 40 years.

To those accustomed to deducing theorems in mathematics from given premises, the method of drawing conclusions from uncertain premises, or generalizing from a sample, which is the subject matter of statistics, might appear as an unsafe game. However, it is the latter that matters in real life and we have to devise ways of using uncertain knowledge to our best advantage. We are taking risks when we choose a partner for life, decide on a particular career, or make an investment. The key to the problem of taking wise decisions under uncertainty lies in quantifying uncertainty and specifying the risk we are willing to take in making a decision. This is the subject matter of statistics, the new discipline conceived and developed in the last century.

On the theoretical side, as the head of the Research and Training School of the ISI, I developed a variety of courses leading to the master's degree with an option to specialize in certain areas of statistics where there is great demand for statistical personnel. I have also developed the Ph.D. program in different areas of theoretical and applied statistics, and personally supervised the work of about 50 students for the Ph.D. degree, who in turn produced about 250 Ph.D.s. Most of my Ph.D. students have made valuable contributions to statistical theory and methods and are occupying high positions in academic and research institutions all over the world.

On the applied side, a major contribution is the establishment of specialized branches of the ISI to help industry in quality control of manufactured goods and development of new products through experimentation using what are called orthogonal arrays developed by me, which *Forbes* magazine has referred to as a *New Mantra* for American industries. The statistical techniques combined with efficient management based on statistical information contributed to production of exportable goods and the rapid economic development of India.

I consider myself very fortunate in that my research work has been recognized through the award of the U.S. national Medal of Science and election as a member of several national science academies and professional societies, and in that some technical terms arising out of my work such as "Cramer-Rao inequality," "Rao-Blackwellization," and "Rao's score test" are incorporated in all textbooks on statistics. The term "Cramer-Rao inequality" is frequently quoted in engineering literature and in some research papers on theoretical physics. Two of my papers are included in the book *Breakthroughs in Statistics*, which deals with the last century. Some of my books are translated into several foreign languages.

The knowledge acquired by statistical means enables us to take wise decisions at individual and institutional levels, detect frauds, settle disputed paternity, use DNA evidence in courts of law, make weather forecasts, determine the efficacy of new drugs, take policy decisions on political and social matters, make better diagnosis of diseases, contradict superstitious beliefs, and so on. As a general advice, if there is a problem to be solved, it is better to seek statistical advice than to appoint a committee of experts. Statistics can throw more light than the combined wisdom of the articulate few.

Roald Hoffmann

Theoretical chemistry. Nobel Prize in Chemistry, 1981, for theories concerning the course of chemical reactions.
Frank H.T. Rhodes Professor of Humane Letters, Cornell University.

I am from the last generation of Hitler's gifts to America. Born as Roald Safran in a happy Jewish family in Zloczow, in southeastern Poland, I survived the war, with my mother. Few others in the family did. I was 11 when my mother and stepfather and I came to New York City.

There was no need to tell us to study. We saw that in America the world was open, and to an outsider, an immigrant, open in just that way, through education. New York City teachers (who, had it not been for the Great Depression, might have been off doing other things) were wonderful. You can imagine there was a good bit of subtle pressure to become a doctor, or otherwise enter a profession. Under the picture of that crewcut boy in the 1955 Stuyvesant High School yearbook, where it says "career aim," I put "medical research." And in high school the only advanced science course I did not take was chemistry.

My path to chemistry was not straight. It took one year of college for me to work up the courage to tell my parents that I didn't want to be a doctor. Meanwhile, I had fantastic teachers in the humanities and arts at Columbia College—Mark Van Doren in poetry, Donald Keene in Japanese literature, Howard McParlin Davis in art history. The world, a world of art and literature, opened up for me.

But I did not have enough courage to go into the humanities. What to do? Looking at my brilliant classmates, I thought I wasn't good enough for physics (I was wrong). Somehow biology did not attract. Summer research experiences in chemistry pulled me in.

So I went to graduate school in chemistry, at Harvard. But even the first two years, I wasn't sure I wanted to be a chemist—I sat in on courses in other departments; I found a graduate student exchange that took me to the Soviet Union for a year. Eventually I found my vocation there, through the mentorship of three men—Martin Gouterman and William Lipscomb first, and then, just after my Ph.D., in an inspirational collaboration with R. B. Woodward, the greatest organic chemist of his time, intellect incarnate.

I came of age as a theoretical chemist in sync with the first computers, which I used with enthusiasm. Then something interesting happened. Instead of succumbing to the psychological traps computer use engenders, I found my way from computing to understanding. The computer spewed out numbers; I learned how to trim away those numbers, so that the closer I got to writing a scientific paper, the fewer numbers there were. I understood, only intuitively then, that the language of chemistry is a mix of symbolic and iconic representations of molecules, of the bonds between atoms, and the shapes that govern properties. I found a way to make small drawings of orbitals, the places where electrons dwell—drawings that were portable, that could be sketched on a piece of paper. And not just by me. Other people did quantum chemistry and calculated orbitals. But I think I found a way to put them into the hands and minds of every chemist.

It was the time of simplification, of finding one reason for a reaction going one way or another. Only later did I learn to appreciate the differences, the rich complexity of the real world.

It has been the greatest fun to make sense of the shapes and reactions of every kind of molecule in the world—from organic, through organometallic, and inorganic, to the borderland of surfaces and solids, where I work now. I see the connections between everything molecular in this world.

And not just to build bridges between parts of chemistry. I loved the English language, the only language I could write in. What excited me at Columbia remained with me. In time, it made sense to try poetry. I thought, naïvely at first, one could just write of the excitement of doing science. But I should have focused on the language, its life-giving tensions.

The language of science is a language under stress. Words are being made to describe things which are indescribable in words, a molecule found for the first time, equations. Words do not, cannot mean all they stand for, yet they are all we have to describe experience. By being a natural language under stress, the language of science is inherently poetic. There is metaphor aplenty in science. Emotions emerge shaped as states of matter, and matter acts out what is in the soul.

One thing is certainly not true, that scientists have some greater insight into the workings of nature than poets. Perhaps we do, but in such carefully circumscribed pieces of the universe! Poetry soars, all around the tangible in deep,dark, through a world we reveal and make. In time I built, and am still building, a land between chemistry, philosophy, and poetry. It is a land in which I need not separate my worlds.

Martin Rees

Formation of cosmic structures. Professor of Cosmology and Astrophysics, University of Cambridge.

I grew up in a Shropshire village—rather remote and beautiful country in the west of England—where my parents were schoolteachers. I can't claim to have had any special infatuation with science during my childhood. I was interested in numbers, and in natural history, but shifted toward mathematics and physics more because I was bad at languages than for any positive reason. However, I was fortunate in my schooling, and gained entry to Trinity College, Cambridge. By the time I graduated, I realized that I wasn't cut out to be a mathematician, so I tried to find a subject where a more synthetic style of thinking was needed—for various extraneous reasons, the choice narrowed down to economics or astrophysics.

I chose astrophysics, which proved a lucky choice for two reasons. First, this was a time (the mid-1960s) when the subject was just opening up. There was, for the first time, genuine evidence for a Big Bang, and perhaps even for black holes. When a subject is new, the experience of older people is at a heavy discount, and it's easier for young people to make a quick mark.

Second, I was fortunate to be in the research group led by Dennis Sciama—an inspiring and charismatic scientist, who had attracted a lively research group. (Stephen Hawking, for instance, had joined it two years before me.)

Over my career, I've worked in many universities in the UK and abroad, but have mainly been based at King's College, Cambridge, which offers a uniquely stimulating and attractive environment. One great advantage of Cambridge is that it's so compact. Each college is a community. I've never felt tempted to defect to the United States, because no American university has these features. And certainly none matches our architecture. One of the joys of being here is just walking through King's College—especially in the evening, when one hears the echoes of organ music from the lighted chapel.

And I've been lucky that astrophysics and cosmology have surged ahead at an exhilarating rate. Although the 1960s were exciting, the rate of discovery has been even greater in recent years. We've discovered that there are planets orbiting hundreds of other stars, we've probed back to the earliest stages of cosmic history, and subjects that were once on the speculative fringe are now part of the mainstream.

From 1860 onward, we have learned from Darwin and the geologists about the history of Earth and the creatures on it. Astronomers and cosmologists are now setting Earth in a cosmic context—mapping out the vast universe, and tracing the origin of stars and atoms right back to a so-called Big Bang nearly 14 billion years ago. Cosmology used to be a subject where there were hardly any firm facts, and speculation had free rein. But we are now inundated with data, and have pinned down some of the key cosmic numbers.

Our work, like that of Darwin, interests a wide public. I'd derive far less satisfaction if it interested only other specialists. We need more people like the late Carl Sagan. It was fortunate that the leading expositor of the space program was someone with his humane eloquence, not a NASA functionary spouting acronyms. But the emphasis has, rightly, shifted toward unmanned exploration. I'm old enough to have grown up thinking of "men in space" as futuristic. But the lunar-landing program was over long before my present students were born. It's just a historical episode to them. But we've all been inspired by the cosmic visions opened up by probes to the planets, and by the Hubble telescope.

I'm often asked what impact the amazing discoveries about the universe have on religion. My answer is a dull one—I don't think they have any distinctive impact. I think there should be peaceful coexistence between science and religion, but there's limited scope for dialogue between them. Among my colleagues, there's a wide variety of religious attitudes. One thing I've learned from science is that even the simplest-seeming things—single atoms, for instance—are hard to understand. I'm therefore skeptical about any dogmatic claims to know the complete truth. But I'm filled with wonder at the complex cosmos we're part of, and that our brains are somehow attuned to make at least some sense of it.

Colin Renfrew

Archaeology. Emeritus Disney Professor of Archaeology, University of Cambridge.

How have we, as human beings, come to be as we are? Every culture, every civilization has its own origin myth, its own story of the creation of humankind. But only prehistoric archaeology can set out to give an answer that is based on hard material evidence. Only prehistoric archaeology can set out actually to investigate the origins of humankind and of human societies by "digging up the past." That is why I am an archaeologist.

The bad news may be that the life of an archaeologist is not really comparable to that of an Indiana Jones: there is less violence, less treasure hunting perhaps. The good news, on the other hand, is that our quest is actually much more interesting than his. Indiana Jones seems to spend much of his time looking for buried treasure, with a high cash value. But if you really have the archaeology bug, you are not much concerned with the cash value of finds—no self-respecting archaeologist wishes actually to keep, still less to sell, the things one digs up. We are looking to discover new knowledge about the past. That is the fascination. To understand how things were and how they changed. The finds, once catalogued and published, belong in a museum where they can be widely informative. Their importance is what they can contribute to our understanding of the human story.

Digging can be romantic and can bring its surprises. There is real pleasure in excavating a bronze age sanctuary in Greece, as I have done, and finding a whole series of small sculptures—cult figures or offerings—to a goddess who was worshiped three thousand years ago. There is a fascination in excavating a stone burial monument in the Orkney Islands, off the north coast of Scotland, and being the first to enter a beautifully built burial chamber that has stood undisturbed since before 3000 BC, one of the oldest surviving buildings in the world. These are good moments.

Even more interesting, however, is to have the opportunity of contributing to some more coherent understanding of how our world came to be the way it is. One of the great advances in the human condition was the transition from the life of the mobile hunter-gatherer to that of the settled farmer, with the beginnings of village life and the origins of agriculture. Another was the urban revolution and the origins of the first cities and the first writing. Both these are stories which we can now begin to tell on the basis of the excavations and the finds. One of my own most interesting experiences was to understand and then to show, on the basis of radiocarbon dating, that the prehistoric monuments of western Europe, the so-called megaliths, are actually the earliest stone-built monuments in the world, older than the pyramids of Egypt. The birth of architecture. That was a literally epoch-making change in our understanding. It showed that the so-called barbarians of prehistoric Europe were creative and ingenious: that lesson has much wider applications for the world at large. Archaeology is relevant in a multicultural world.

Miriam Rothschild

Small animals, botany. Independent researcher.

Naturalists are born and not made. I loved insects, particularly ladybirds, which I began to collect at the age of five years old, but my development into a so-called scientist was due entirely to the influence of my father, who was himself a first-class scientist and who discovered the flea vector of plague. He studied fleas and butterflies in his spare time, although he was a full-time banker.

In our home, natural history was not a subject—it was a way of life. My father was also very gifted in the way he treated his children. For instance, when he himself went out collecting material, whether it was plants or caterpillars, he took me along as well and treated me as if I were a contemporary, not a child playing with toys. As early as the age of five or six, I was counting the spots on the forewing of ladybirds, and could already tell the difference between the small tortoiseshell butterfly and comma.

Any group of animals I happened to come into contact with, I have always wanted to study. When visiting the Marine Biological Association at Plymouth, I happened to see a living larval trematode worm through a microscope, and I was so fascinated by the attractive larval stages of this parasite that I began to study its life cycles, and continued to do so for the next 10 years. One of the discoveries that I made concerned the influence of parasites on the growth of the intermediate snail host. These developing worms produced gigantism in their host, which grew faster and larger than nonparasitized individuals. I then transferred to the study of agricultural pests, which, fortunately for me, included the study of the wood pigeon and its parasites. This led to my publication of a monograph on fleas (now Siphonaptera), which included the discovery of the mechanism of their jump, and perhaps an even more interesting fact was that in order to breed, the female flea appropriates the sex hormone of the mammalian host. In the earliest trials, carried out by Professor Harris and myself on the effect of the human birth control chemical, we tried its effect on the rabbit flea feeding on the ears of a rabbit injected with clormadinone and found that the fleas were completely sterilized.

Unfortunately, this brilliant collaborator died before he was awarded the Nobel Prize. I then abandoned fleas and returned to an earlier interest in Lepidoptera (moths and butterflies), which has endured to the present day. This fascinating and exquisite group of insects provided poetic and aesthetic pleasure, as well as chemicals of medical importance. They have shown that conservation of wildlife, plants and animals alike, can be of vital importance to humankind. This is emphasized by the discovery of heart poisons in the monarch butterfly by myself and Professor Reichstein, and a potentially powerful antibiotic, which may eventually save many thousands of human lives.

The study of butterflies is, in a sense, the gateway to the entire natural world, and it can bring great satisfaction to all those who are lucky enough to be born with the necessary gene for scientific curiosity.

Allan Sandage

Astronomical origins and evolution. Astronomer, Carnegie Institution of Washington.

I have come to believe that I knew at the age of 11 what I wanted to do in the years that would come with adulthood. Living at the time in Upper Darby, near Philadelphia, Pennsylvania, I looked through a boyhood friend's telescope at planets and stars, and that opened for me the regularity of the natural world that is independent of the often unstable world of everyday life. The impersonal, rational, regular natural laws of the "outside" seemed to have within them the promise to find truth—truth defined as what actually *is*, not what people say what is. The evident order of nature suggested that one could discover that kind of truth. I became a scientist as an astronomer.

My strong interest in science and mathematics grew during my middle and high school years in southern Ohio. I found that I was fairly good at working the problems in ninth-grade geometry and tenth-grade algebra, and that gaining solutions to ever more difficult problems gave an almost ineffable satisfaction. The evident truth reached, for example, in the formal proofs of Euclidean geometry showed that problem solving in science could lead to a life of fulfillment and purpose.

After two years in physics at Miami University in Oxford, Ohio (my hometown), 18 months in the navy during the Second World War, two years in physics at the University of Illinois, and five years of graduate work in physics and astronomy at the California Institute of Technology, I became a staff astronomer at the Mount Wilson and Palomar Observatories.

With continual access to the world's three largest telescopes in their times (the Mount Wilson 60- and 100-inch reflectors and the 200-inch Palomar Hale titanic), my career as an observational astronomer began in 1950.

Along the way, with many colleagues and companions in the adventure, we would discover the elements of stellar evolution, learn how to age date the stars, uncover quasars that are the parents of black holes in the centers of galaxies, and help find the age, shape, and size of the universe.

From the backyard of Bruce Olsen in Upper Darby in 1936 to the mountaintop observatories in California, Arizona, South Africa, Australia, and South America, I joined the quest to find what *is* with many other observers of that outside, regular, rational, impersonal universe that still promises to reveal objective truth to those who would seek it.

Frederick Sanger

Retired biochemical researcher. Nobel Prize in Chemistry, 1958, for work on the structure of proteins, especially that of insulin. Nobel Prize in Chemistry, 1980, for contributions concerning the determination of base sequences in nucleic acids.

My father was a doctor. He did research on antibodies for a short time, differentiating human and animal blood. He was bright. I'm not one of these intellectual geniuses. I didn't get scholarships. When I first came up to Cambridge, I had to think about what subjects I was going to take. I had planned to take chemistry and physics but had heard about a new discipline of explaining biology in terms of chemistry. I had an enthusiastic supervisor who persuaded me to study it.

Biochemistry is the study of the actual chemical mechanisms which take place in the body and determine how living matter works. Biology and chemistry used to be separate. Little was known about the chemistry and composition of living matter in 1940. Cambridge had one of the first departments teaching the subject.

Living matter is largely made of proteins, which are its active components. The other important components are the nucleic acids, DNA and RNA. And they are essentially the working parts. The DNA is probably the most important component. It contains all the instructions for the making of the proteins and for the functioning of living matter. That wasn't known when I started.

I started working on proteins. They are long chains of smaller units called amino acids. My initial work was to develop methods for working out the arrangement or sequence of amino acids in the protein. The function and activity of proteins depends on this order. The proteins have up to 1,000 amino acids. The insulin molecule is a small protein, having only 50 amino acids. It was the first protein whose structure I determined, and it is for essentially this work that I was awarded my first Nobel Prize. The sequence of proteins is determined by the sequence of the DNA, which is built up of four units, the nucleotides.

After working on proteins and determining the structure of the insulin molecule, I changed over to study the sequence of the nucleic acids (RNA and DNA). These are much larger molecules, and different methods had to be developed. The DNA has only four different components. You can imagine a book of instructions with four different letters. It's limited, but it manages to make living matter. I was prepared to take on the more difficult experiments and could do so because I already had the first Nobel. I could afford to not get results for a year or two and still have a job. There was very little competition in the field I was working in because the molecules were so large. I succeeded in devising a method for determining these sequences, and for this I was awarded a second Nobel Prize.

The DNA of humans contains about three billion subunits, and after I retired, this sequence was almost determined. The scientists now are essentially using the same method which I developed in the lab with only a few people, but now it is much more automated. At present, the details of how the DNA does function and how it can affect human beings are being worked out, mostly in private companies. DNA sequence is characteristic for each person, and it varies slightly. One may find out that a certain change causes a certain disease and realize that change needs to be amended. That's not so easy. That's the holy grail.

I like messing about in the lab, doing experiments, thinking them out, working things out for myself. It's absorbing work because you're always thinking about it. It's exciting. It can also be fairly frustrating because you're doing things that haven't been done before. Most things you try don't work out and some people get frustrated. I found the best thing to do when an experiment didn't work was to forget about it and start the next experiment. It keeps you on your toes.

The work you do depends on more than just how bright you are. Some people are too bright and know all the answers. They're impatient. I was fascinated by what we're made of and didn't need much motivation beyond that. I spent most of my time thinking about going from one experiment to the next. I like using my hands. I wasn't working on a grand design. I had a medical interest behind my work. I wanted to see if I could help.

Maarten Schmidt

Evolution and distribution of quasars. Professor of Astronomy, Emeritus, California Institute of Technology.

As a youngster, I first wanted to become an accountant like my father. He often in winter nights would take me for a walk in the outskirts of Groningen. After May 1940, in occupied Holland, there was a blackout every night and the starry sky would be brilliant, if we were not interrupted by Allied bombers and antiaircraft guns. In 1942, my uncle who was a pharmacist and amateur astronomer introduced me to the planets and stars with his telescope. That was to be the beginning of my lifelong interest in astronomy.

I have spent most of my career at Caltech, where the telescopes, especially the 200-inch telescope at Palomar, were a major asset. Quasars were definitely the most exciting objects I have worked on. Once their large redshifts were found in 1963, it was clear that they should be visible out to enormous distances. Eventually, in the early 1980s, I set out with Jim Gunn of Princeton University and Don Schneider, then my postdoc at Caltech, to search systematically for the most distant quasars. This involved searching for quasars with the 200-inch for whole nights, then carrying out follow-up spectroscopic observations to determine their redshifts. We found more than a hundred quasars with large redshifts, the largest being 4.9. In the spectrum of this quasar, all the spectral lines are observed at a wavelength 5.9 times the laboratory wavelength.

Since redshift in the expanding universe is a distance indicator, this was at the time the most distant known quasar: its light left on its way to us almost 12 billion years ago, only a billion years after the Big Bang. The main scientific result from our work was that quasars grew in numbers from 1 billion to 2.5 billion years after the Big Bang, after which, according to earlier results, they started to decline till the present day.

Besides some applications of imaging analysis in medical technology, material benefits from astronomy are rare. What, then, moves hundreds of astronomers every clear night to work hard and long at the telescope, sometimes in bitter cold, to do their observations? Their motivation must be part of the human urge to explore the environment. After all, the universe itself is the ultimate environment. We are drawn to our origin and to the early stages of the universe when it all started. As astronomers, we are a community, working as it were on a tapestry of nature that is slowly taking form. If we are lucky, we work on the middle of the tapestry. When we age, we work on the periphery.

John Schwarz

Superstring theory. Harold Brown Professor of Theoretical Physics, California Institute of Technology.

My parents, both of whom were scientists, were very supportive of my early interest in math and science. During my undergraduate years at Harvard, I majored in math, which I thoroughly enjoyed. However, when it came time to apply for graduate school, I decided to switch to theoretical physics. I wanted my mathematical formulas to have something to do with reality. At Berkeley, where I was a graduate student, my tastes in physics were strongly influenced by my adviser Geoffrey Chew. After receiving my Ph.D. in 1966, I had a six-year junior faculty stint in Princeton. Then I moved to Caltech, where I have remained ever since. Until 15 years ago, the Caltech particle theory group featured Richard Feynman and Murray Gell-Mann (both Nobel laureates), who also influenced me greatly.

Almost all of my career as a theoretical physicist has been devoted to the study of string theory. This is a type of relativistic quantum theory based on fundamental objects that are tiny loops (called strings) rather than points, as is the case in more conventional quantum theories. String theory arose in the late 1960s as a candidate theory of the strong nuclear force. This is a force that holds neutrons and protons together inside the nucleus of an atom. However, in the early 1970s a very successful alternative theory of the strong nuclear force, called quantum chromodynamics (or QCD), was developed. As a result, most physicists stopped working on string theory. However, I was so enthralled by the mathematical beauty of string theory that I continued to work on it. This persistence paid off when, in 1974, the late French physicist Joël Scherk and I proposed to change the goal of string theory to one that is much more ambitious: the construction of a unified quantum theory containing gravity and all other fundamental forces.

An important ingredient of string theory is a type of symmetry, called supersymmetry, which relates those particles responsible for stable matter (fermions) to those that transmit forces (bosons). As a result, the subject has been dubbed superstring theory. If successful, superstring theory should account for the properties of all the elementary particles as well as the physics that controls the origin and evolution of the universe. The basic idea is that the elementary particles correspond to different motions of the fundamental string. Surprisingly, the mathematical consistency of such a theory requires the existence of gravity. As Einstein taught us, gravity is determined by the geometry of space and time. In the case of string theory, it turns out that in addition to the three familiar dimensions of space, six or seven extra spatial dimensions are required for mathematical consistency. They should form a tiny space (attached to every point in ordinary space) that is too small to have been observed. The details of the geometry of the extra dimensions play an important role in determining the physics that is observed.

Major advances in understanding aspects of superstring theory that apply when quantum effects are small were made in 1984–85 (the first superstring revolution). Since 1995–97 (the second superstring revolution) and in subsequent years, there has been a dramatic shift in understanding superstring theory when quantum effects are large. One result is that what previously appeared to be five distinct theories are now understood to be five different limits of a unique underlying theory. Despite all that has been achieved, superstring theory is still a work in progress that is far from completion. Many very clever people are working on it with great dedication, and so progress continues to be rapid.

In the most likely scenario, the distances and energies at which stringy effects are observable are very far from what can be observed with current technology. This makes it difficult to construct decisive experimental tests of string theory. However, there are a few possibilities. One, which many people are exploring, is that the distance and energy scales are not as extreme as is usually assumed, so that they could be observed in future accelerator experiments. It would be great if that were the case, but I am skeptical. A more likely possibility is that the supersymmetry partners of the known elementary particles will be observed at future accelerators, such as the LHC in Geneva, Switzerland, which is scheduled for completion in 2007. This discovery would not prove that string theory is correct, but it would be extremely encouraging and informative. The lightest supersymmetry particle (LSP) is a leading candidate for the dark matter that composes about 80 percent of the mass in the universe. Experimental efforts to detect this dark matter are under way.

What is the relevance of this work to lay people? In my opinion, most important is the satisfaction that comes from knowing that humankind is making great strides toward unraveling the deepest mysteries of the physical universe. Isn't that enough?

Steven Stanley

Paleontology. Professor of Paleobiology, Johns Hopkins University.

I am a paleontologist—a student of the history of life and environments on Earth. My enthusiasm for this subject is not only intellectual but also aesthetic and romantic. I am excited by the rising up of great mountain chains, the waxing and waning of glaciers, the evolution of novel creatures, and the sweep of mass extinctions.

Creativity is everything to me. Thus, as a student, I completed three theses without advice from teachers: my senior thesis in high school on the geological history of the valley where I lived, an undergraduate thesis on a fossil coral reef, and a doctoral dissertation on the functioning of mollusk shells of various shapes. In college, I learned I could do scientific research without assistance, and it became clear that only an academic career would allow for unfettered investigation of whatever phenomena cried out for explanation.

Ironically, I had always hated school, except for rare creative projects. At the age of 46, I learned why. Cognitive tests revealed that, from birth, I had lived with attention deficit disorder, a condition that impairs reading and memorization and induces carelessness but is often associated with inventiveness. Previously, my life had made no sense. I was much better at doing science than learning about it. Only through painfully hard work had I done fairly well in school. Curiously, having always struggled in the classroom, I had at the age of 32 become the youngest full professor at Johns Hopkins University.

In the early years, my father unwittingly did me a great favor. Though a gifted man, he too had disliked school. He had no attention deficit but was in some way dyslexic: a college graduate unable to spell common words. Having been frustrated in the classroom, he never valued grades. Generating original ideas was his passion. Among other things, he invented a machine that polished all the ball bearings made in North America. When it came to schoolwork, he let me off the hook, frequently declaring creativity superior to rote learning.

Occasionally, I point to a small book on the top shelf of my office called *The First Book of Stones.* This slender volume got me started, at the age of nine. Minerals became my obsession. They were beautiful. Fossils seemed drab by comparison; they were finally brought to life for me by a delightful professor at Princeton named Al Fischer. At the same time, I became excited about organic evolution, through an introductory biology course in which, while the trivia of exams condemned me to mediocre grades, the lectures were utterly inspiring.

Although I usually call myself a paleontologist, in the language of the 21st century, I am a paleobiologist or geobiologist. These new labels are not simply buzzwords. They depict two transformations—one might say renaissances—in my field, both of which I have been privileged to take part in. The first made the study of fossils more biological. The second, still in progress, is about linking the history of life to past environmental change—to shifts of climate and ocean chemistry, for example. This work is not purely academic. It sheds light on the future of our fragile planet and its vulnerable inhabitants. I have, for example, published a theory on the cause of the modern ice age that yields an admonition: if, as I think it may, human-induced global warming destabilizes the Arctic Ocean, the world will plunge into a frighteningly balmy state. In effect, we may find ourselves in a pre–ice age world, with much warmer temperatures and higher seas than are predicted in conventional scenarios of future global warming.

Joan Steitz

Structure and function of small RNA-protein complexes. Professor of Molecular Biophysics and Biochemistry, Yale University.

Science is full of surprises.

I count myself privileged to be a part of the 20th-century revolution in biology. As a college chemistry major in 1961, I was fortunate to learn about the discovery of the double-stranded structure of DNA—even before it appeared in textbooks or was the subject of courses. Its simplicity immediately engaged me. I was fascinated by the possibility that there might be a molecular explanation for the genetic phenomena that I had found so intriguing in high school.

The field of molecular biology that I subsequently entered as a graduate student in 1963 was small and eclectic. It focused exclusively on simple systems such as bacteria and their viruses. The idea was that molecular understanding of life could be achieved only by starting with the most uncomplicated organisms. Investigators bold enough to work on higher cells were even derided as wasting their time on something that was too complex to be understood.

It was therefore unimaginable what would happen in biology within my lifetime. Although I watched the step-by-step progression, I still find it amazing that we now know the sequence of the four billion base pairs of the human genome. In contrast, one of my co–graduate students in 1968 wrote his entire thesis on determining the sequence of a single base (at one end of a bacterial virus genome)! Also unimaginable were the far-reaching implications the revolution in biology would have—for medicine, in spawning a multibillion-dollar biotech industry, even for forensics.

It was likewise unimaginable that women would hold important positions in science and academia. When I was a graduate student, there were no women professors at major research universities. Women expected, as I did, to become research associates working in the laboratory of a kindly male mentor. Today, I am fortunate to have women colleagues who face challenges similar to mine at universities both in the United States and around the world.

Other surprises have included how important serendipity can be in the pursuit of science. For example, it was the chance comments of another scientist that led us to begin working a year later with the sera of patients with rheumatic disease (such as lupus). This, in turn, led to the discovery of a new class of particles inside cells that are essential for pruning out the nonsense segments that interrupt our genes. These particles, commonly known as snurps, allow the information contained in genes to be converted into the working proteins of a cell. This is a relatively rare example of the biomedical paradigm flowing backwards: from bedside to bench. Tools known to physicians proved to be invaluable for gaining an understanding of one of the most basic processes occurring inside cells, that whereby DNA makes RNA and RNA makes protein.

The central role played by RNA molecules in all cells is yet another surprise. When I began, three classes of RNAs were known. The messenger RNA (mRNA) takes the information in the DNA to the protein synthesis machinery. The ribosomes that make proteins are largely RNA (rRNA). And transfer RNA (tRNA) brings the amino acids to the ribosome to be strung together into proteins. But now there are two additional classes. One is the small RNAs (snRNAs) that are part of the snurps we characterized and are essential for the splicing process. Only very recently, microRNAs have emerged as critical regulators of which proteins are made. All these discoveries underscore the importance of RNA in the functioning of all cells and strengthen our belief that RNA preceded DNA in the evolution of life processes.

A final surprise is that it is almost as much fun to share the joy of discovery with a talented younger colleague as to make the discovery oneself. I have been privileged to attract to my laboratory a number of extremely bright and creative students, at the undergraduate and graduate as well as at postdoctoral levels. They, along with my mentors, colleagues, and family, are responsible for my picture's appearing in this collection.

Lynn Sykes

Tectonics, nuclear arms control.
Higgins Professor of Earth and Environmental Sciences, Lamont-Doherty Earth Observatory, Columbia University.

When I was 10, my aunt Ethel sent me a chemistry set for Christmas. Contrary to my mother's dire warnings, I neither poisoned myself nor blew up the house in Arlington, Virginia. It did kindle my interests in science. A stamp from Trinidad on a letter to my father when I was eight led to stamp collecting and a lifelong interest in geography and history. We moved to then rural Fairfax County when I was 12. I helped my father, an air traffic controller who had started with the Weather Bureau in 1929, build our house.

I owe much to the U.S. educational system. I flourished in science and math at high school. Mr. Taylor, a former official in the U.S. Department of Agriculture, took special interest in me in his classes in world history and American government. I was fortunate to receive a full scholarship to the Massachusetts Institute of Technology. A freshman course persuaded me to major in geophysics and geology. My years there also awakened interest in the humanities, especially the history of philosophical thought.

As an undergraduate, I was told bright scientists should not waste time on scatterbrained ideas like continental drift. Arriving at Columbia's Lamont Geological Observatory as a graduate student in 1960, I discovered there the same attitudes that existed at most American universities. Nevertheless, Lamont was abuzz with new and exciting data collected by seismographs from around the world and vessels that sailed the oceans. For my Ph.D. thesis, I examined thousands of earthquake records made by Lamont instruments. My work involved seismic waves that propagate in the crust of the Earth beneath oceans. Earthquakes in remote oceanic areas that created the waves I was studying were very poorly located. Needing better locations and origin times, I used a computer program, one of the first of its kind, to obtain more accurate locations of hundreds of earthquakes along mid-ocean ridges. When Maurice Ewing and Bruce Heezen of Lamont traced the ridge system from the Atlantic into the Indian and Pacific oceans a few years earlier, it was clear that it was one of the major geological features of the globe, dwarfing the Alps and Rockies in its extent. With more accurate locations, I discovered several places where earthquakes suddenly changed direction for hundreds of kilometers only to zig back again and follow ridge crests. I quickly realized that undiscovered geological features such as these in the oceans were more important than the waves I was studying, wisely deciding to work on them full time after finishing my thesis.

In 1965, Tuzo Wilson of Toronto University proposed that my zigzag pattern of earthquakes was indicative of a new class of great geological features, which he called transform faults. His hypothesis required that new seafloor is being added along ridges, causing continents to move. I realized I could test Wilson's theory by using seismic waves. Stopping work on what I was doing, I immediately plunged into a test of his speculative hypothesis. Within a few months, my data confirmed Wilson's theory.

Soon a more inclusive theory called plate tectonics was proposed, whereby not only continents move but also that seafloor, created at mid-ocean ridges, returns to the Earth's interior near deep-sea trenches like those off Japan and the Aleutians. During 1968, I collaborated with the Lamont seismologists Jack Oliver and Bryan Isacks in pulling together a large body of seismological data to confirm the theory of plate tectonics.

My work on plate tectonics has led me into two additional fields—the quest for a verifiable ban on nuclear testing and earthquake prediction on time scales of a few decades. In 1959, seismology suddenly changed from an obscure, poorly funded field because it then provided tools for detecting and identifying underground nuclear explosions. I demonstrated that plate tectonics provided an explanation of why earthquakes occur where they do and why certain seismic waves propagate efficiently in some geological terrains.

In 1974, I was a member of the U.S. team in Moscow that negotiated the Threshold Test Ban Treaty, a bilateral agreement that set an upper limit on the sizes of underground atomic tests. That experience reinforced an earlier realization that improvements in the science of verification were not enough to end nuclear testing. The Comprehensive Test Ban Treaty, which was signed in 1996, continues to be opposed by many in the United States. I have stated repeatedly in Senate testimony, scientific papers, and op-ed articles over the past 30 years that it is highly verifiable and very much in our national security interest. Experiences going back to my childhood near Washington nourished in me the concept of science in the public interest and the notion that one individual can make an impact on issues of great importance. It is my belief that few are as important as the prevention of nuclear war.

Sir John Sulston

Gene function. Nobel Prize in Physiology or Medicine, 2002, for discoveries concerning genetic regulation of organ development and programmed cell death. Researcher and Former Director, Wellcome Trust Sanger Institute, Cambridge, UK.

I feel I always was a scientist. For me, science is very much a matter of curiosity and finding out. I sometimes call it "playing with the toys," though you're not supposed to say that. It's supposed to be very serious and important. But, actually, it's about playing with the toys. When you play with toys enough, you start to do things that are different, and you suddenly discover you are seeing something that people haven't seen before, and you just carry on that way. Of course, at the same time you are trying to understand something, and according to how good you are (I put myself very much in the middle rank on this), you can synthesize it more or less into some new concept.

My work on the function of genes began with Sydney Brenner's nematode worm *Caenorhabditis elegans*. We discovered some of the functionality by knocking genes out chemically and then looking for funny worms, worms that couldn't move properly or were an odd shape. Then you knew that the gene you'd knocked out had to do with that function. The problem was getting back into the DNA to find it. The DNA, which is the repository of genetic information, in the case of the nematode consists of 100 million bases—a 100 million letter code. At that point, although Fred Sanger had got his sequencing method going, it was a huge job to actually find the genes, because of the quantity. We didn't have today's automated methods to handle all of it. The first thing I did in the early 1980s was to start a mapping project to more quickly discover where the genes were. That took us through the 1980s. With improved technology, we were then able to sequence the entire genome.

By this time, I was in collaboration with Bob Waterston. We were both driven by the utilitarian aspect. We weren't sequencing for its own sake. We were sequencing because we knew that was the way to find out more about how the worm worked. As we were forging ahead with the worm, we suddenly found ourselves in the position of temporarily having the biggest sequencing labs in the world. We were drawn into the Human Genome Project to get that done too. It wasn't a holy grail for me or for Bob, but an absolutely pragmatic, step-by-step approach because this would then allow us to find out how the whole thing worked. Regarding this work, I don't like it scientifically when people make too much of the human genome, because it isn't a great breakthrough in itself. It's just the natural outcome of everything that was going on before. I do like it in a social sense because the excitement about it and the potential has opened the dialogue. I am happy that biology has come out of the closet.

The international groups that worked on the human genome have more to do, two things in particular. One is to sequence a lot of other genomes. That is happening now, faster and faster because it is increasingly less expensive. It's incredibly useful. We had the worm in the beginning, then the fruit fly, and then the mouse, farm animals, and plants. Comparing all these genomes as well as using them individually to find out about how the particular organisms work is very valuable.

The second objective, which is being worked on very strongly, is the study of human variation. Here we get into something of immediate value, to correlate human variations with medical conditions because it allows one to make more accurate diagnoses not only about hereditary disease per se but also about our propensity to other problems like diabetes and heart disease. These variations can be very valuable as well as teaching us about which genes are important in building the body in various ways.

There is a social side to this that is enormously important and something I spend most of my time on now. It is to ensure equity of treatment between individuals in the use of this genetic information. Also, we must use this information globally. There's a great tendency at the moment for market-based health care to be feeding lots and lots of new developments into the northern rich countries, and a huge gap is opening between them and the developing countries. We have to fix that. People have different views about how to fix it. Personally, I think we should take affirmative action rather than let things drift on in a market-based way. Very specifically, people should not think that from now on all worthwhile biology is big biology. People need to use the information to look at the individual biological systems and assemble them. Because of the great success and importance of being able to decipher masses of information in the human genome, there's a tendency among people to think we have to go on doing this kind of thing—big machines, lots of data. There's always a place for that, but it should be regarded as a tool, not as an end in itself.

Ian Tattersall

Paleoanthropology. Curator, Department of Anthropology, American Museum of Natural History.

There are many reasons for becoming a scientist. I became one by accident, stumbling with great good fortune into human evolutionary biology in the course of a rather unfocused attempt to become educated. But once there, what as much as anything else kept me in science was that scientists, unlike professionals of many other kinds, don't always have to worry about being right. Many people assume that science is a rather absolutist affair that progresses by piling up objective knowledge for the ages. Far from it! Science is a constantly self-correcting process in which notions are advanced in testable terms and are then discarded if found wanting. What makes an idea scientific is not that it has been shown to be true but that it is expressed in such a way that it can at least potentially be proven to be false. And this is what makes science fun. Because scientists can let their imaginations run rampant, while their feet are still kept firmly on the ground by the knowledge that their colleagues will sooner or later catch up with any flaws in their arguments. Science may have its stars, but in the end its communal nature dominates, and all scientists can take satisfaction in participating in a great majestic process, flowing through time, that is far bigger than any one of them.

My special interest at the moment is in the rather intractable matter of exactly how humans acquired their intellectual uniqueness. How did our precursors transition from an ancestral cognitive condition that was nonlinguistic and nonsymbolic, to give rise to the linguistic, symbolically reasoning creatures that we are today? Clearly, there is a gulf here: our unprecedented capacities are not just an extrapolation of earlier trends. And to explain how this gulf was crossed, we have to appeal to the notion of emergence, whereby a chance coincidence of acquisitions produced something truly novel—and something, moreover, that had nothing whatever to do with adaptation. Some scientists believe that natural selection has honed us over the eons to behave in particular ways: that, behaviorally, we *Homo sapiens* are in some sense prisoners of our genes. And the search is on for "the gene" for everything from homosexuality to infidelity. Quests of this kind are just the sort of thing that makes science dynamic, opening up new alleyways for investigation. In this case, though, I think the alley is blind. To me it is as clear as anything in science can be that the human species has not been fine-tuned by evolution for anything. As individuals, our natures are doubtless greatly influenced by our genes; but as a species *Homo sapiens* is gloriously (or appallingly) paradoxical, embracing any pair of behavioral antitheses you'd care to mention. And whatever side of the scientific debate you happen to be on, it is with this biological reality that our social policies will eventually have to come to grips.

Harold Varmus

Cell biology. Nobel Prize in Physiology or Medicine, 1989, for discovery of the cellular origin of retroviral oncogenes.
President, Memorial Sloan-Kettering Cancer Center.

In his last collection of essays, the naturalist Stephen Jay Gould mentions a senior colleague who says that he continued to do research during his advanced years because its pleasures were uniquely akin to "continual orgasm."

As someone who continues to manage a laboratory (if not actually to do experiments myself) in my early sixties, I can say with some authority that Gould's colleague has got the wrong metaphor. A long-distance sport, say bicycling, is much closer to the experience. Long flat intervals. Steep, sweaty, even competitive climbs. An occasional cresting of a mountain pass, with the triumphal downhill coast. Always work. Sometimes pain. Rare exhilaration. Delicious fatigue and well-earned rests.

For the best scientists I know, the trip is a long one, usually beginning with an overly ambitious itinerary, inspired by goals that are at once commendable, audacious, public-spirited, and self-serving. In my own career, for over two decades virtually everything my lab pursued was influenced by aspirations to answer two overriding questions: How do retroviruses multiply? And by what means do they cause cancers? In the past decade, our sights have focused on a single question, as we have tried to build on what our group and many others learned in the preceding two decades: How do the genes that retroviruses commandeer to cause cancer in animals participate in the development of cancer?

These broad questions are not difficult to position on a map of human knowledge. But it is more difficult—and actually more interesting to most scientists—to chart the particular course of inquiry that is intended to answer them. The route entails the biological systems to be studied, the specific experiments to be undertaken, the technical methods to be used. The information required to plot this trip involves much more than a low-resolution map of the world; it requires local maps that reveal the details of the terrain to be crossed, with indications of safe and dangerous routes, the paths under construction, the degrees of incline, the unbreachable obstacles. Of course, such knowledge is generally unavailable when the excursion begins, and is revealed only by the trip itself, part of the process of discovery.

When I started my own journey in the late 1960s, physician-scientists of the sort I hoped to become were just beginning to realize that the beautiful molecular and genetic methods used to understand the behavior of bacteria and their viruses were ready to be applied to animals (including man) and their diseases. The lure of studying cancer with these powerful new tools was especially great because certain laboratory animals, like mice and chickens, developed cancer if infected by quite simple viruses, harboring no more than a few genes. Cancer itself had a special allure because it is so common, so devastating, and so mysterious—with the suggestion that its cardinal feature, unrestrained cell growth, arose from a distortion of those treasured mechanisms that normally allow us, as aggregates of specialized cells, to become the miraculous creatures we are.

Among the best moments—I must add: thus far—on my own tour through this territory have been a couple of ascents that led to dramatic vistas: one that allowed us to see how retroviruses might capture important genes and turn them into cancer genes, another that showed how retroviruses can perturb such genes to induce a cancer. Both viewpoints turned out to reveal important features of the landscape of cancer—and also new, unconquered heights. As in cycling, in science there is always another hill to climb.

Robert Weinberg

Molecular basis of cell transformation.
Daniel K. Ludwig and American Cancer Society Professor for Cancer Research, Massachusetts Institute of Technology.

I grew up in a family of refugees from Hitler's Germany. We were Europeans who somehow landed in the unfamiliar surroundings of western Pennsylvania. German was the language at home and of almost all my parents' and grandparents' friends. My father's family had been horse and cattle traders in small Westphalian villages, and when the times became difficult in the middle and late 1930s, many of our relatives scattered to the four corners of the earth. Those who, for one or another reason, stayed behind perished, almost without exception, in the camps.

I was impressed, more than anything, with how unpredictable life can be, and that hard work and long-term planning can often be undone by forces well beyond one's control. While growing up, I came to believe, as I do to this day, that luck is a major ingredient in how one fares on this earth. My family's experiences bred in me, for better or worse, a certain fatalism and a cynicism about the forces of good and evil, and which would prevail in the long run.

All this may explain why I have never counted much on long-term planning and setting goals for myself. Instead, I've put one foot forward after the next, hoping all the while not to step in a deep hole. A career in science was never a life goal of mine. To be sure, our family life was permeated with a deep respect, indeed reverence for intellect and learning, but that in itself hardly led uniquely down a path toward science. There were, after all, many kinds of ways to be smart. My father wanted me to take over his dental practice in Pittsburgh—a prospect that did not excite me. Instead, like many young Jewish men at the time, I imagined a career in medicine for myself.

I went off to MIT, in Cambridge, Massachusetts, simply because friends of the family had done so. I knew little about colleges—my parents even less so. Most everyone there was interested in physics or chemistry or electrical engineering, all of which I found to be sterile and uninteresting. Instead, I became excited by the new molecular biology that was then being born at MIT and elsewhere. A decade after the Watson-Crick discovery of the DNA double helix, the revolution in modern biology was in full swing. The prevailing agenda was clear and unambiguous: those who were clever enough could take apart complex biological phenomena and ultimately explain them in terms of DNA and RNA and protein molecules. Cells and tissues were nothing more than complex versions of the Erector sets that I had spent my childhood assembling and taking apart.

A tortuous experimental path led me, finally in the mid-1970s, to enter into studying cancer, at the time viewed as a disease that was hopelessly complex and thus far beyond the explanatory powers of molecular biology. Nonetheless, with the passage of time, the community of cancer researchers began to tease out some of cancer's dark secrets, notably the damaged genes that lie within cancer cells and are responsible for their runaway proliferation. My own group found oncogenes within human cancer cells and discovered that they were damaged versions of normal cellular genes, and so we could formulate a simple scheme of how cancer arises: normal genes instruct cells when they should grow and divide. Carcinogens invade our cells and damage their DNA and thus their genes, and the resulting mutant genes proceed to tell cells to grow uncontrollably. This scheme is simplistic but useful, and it continues to define the paradigm driving contemporary cancer research.

In the mid-1970s, we knew almost nothing of cancer's origins; by now, we understand it in intimate molecular detail. By finding the damaged genes and proteins within cancer cells, cancer researchers can now find new ways to treat disease, since new types of therapeutic drugs can be dispatched to attack these damaged molecules and reverse the course of cancer. This strategy is still more of a hope than a reality, but it will surely yield grand successes over the next generation.

It is certainly the case that many in my area of research have been motivated by the lofty goal of improving the human condition. My own drive has been quite different. From a very early age, I have loved to take complicated things apart and see how they work. I still do.

James Watson

Biochemistry. Nobel Prize in Physiology or Medicine, 1962, for discoveries concerning the molecular structure of nucleic acids and its significance for information transfer in living material. Chancellor, Cold Spring Harbor Laboratory.

I am a scientist in large part because I was born highly curious. Like most of my scientific colleagues, I am a product of the 18th-century Enlightenment that stopped accepting truths from religious revelations and wanted instead explanations based on the newly emerging laws of physics and chemistry. One of the most important pluses for my future was that my father was strongly antireligious. To him life wasn't a product of God but somehow a product of atoms. Francis Crick's youth was not that straightforward. He was raised in an English "nonconformist" (Congregational) family that went to church on Sunday—a day on which fun was not proper. But by early adolescence he also saw no reason to believe in a personal God. That certainly helped our getting along so well. When I first heard Francis talk, I imagined myself hearing George Bernard Shaw expounding on rationality.

I became interested in DNA because I wanted to know what life was. Even after I entered college, biology was not yet in any way explicable in terms of the laws of physics and chemistry. There was the gene, but we didn't know how it could carry genetic information. The 1953 discovery of the DNA double helix let us immediately know how genetic information is stored. The double helix also revealed how genetic information is copied. Through separating its two strands, the information of parental strands is used to lay down the information of the new daughter strands with complementary sequences. When we found the double helix, we solved two big problems—what is genetic information, and how is it copied?

What we didn't know (the third big question at the time) was how cells read genetic messages. Just knowing the structure of DNA wasn't sufficient. We had to discover the cellular machinery that reads the genetic information of DNA. In doing so, we learned that the genetic information of DNA becomes copied into RNA chains of complementary sequences. These, in turn, are used as informational molecules to direct the laying down of polypeptide chains of proteins. This exciting adventure story lasted 13 years, leading to the 1966 establishment of the genetic code.

Steven Weinberg

Particle physics. Nobel Prize in Physics, 1979, for contributions to the theory of the unified weak and electromagnetic interaction between elementary particles, including, inter alia, the prediction of the weak neutral current. Professor of Physics, University of Texas, Austin.

My start in science came with a hand-me-down chemistry set from an older cousin. I enjoyed playing with stinks and bangs, and having those handsome-looking pieces of chemical glassware on my shelf. Then I started to read about chemistry, and I learned that all these chemicals were composed of atoms. To understand why chemicals behave the way they did, you had to understand atoms. I started to read popular scientific books about chemistry and atoms, and as I read, I learned, particularly from books by George Gamow, that there is a very arcane, esoteric branch of science, quantum mechanics. Quantum mechanics is peculiar. It tells you, for instance, that you can't talk about an electron having a definite position and at the same time a definite speed: if its position is definite, you can't know anything about its speed. I found that absolutely mysterious and disturbing. I decided that if this is what you have to learn to understand chemistry, and atoms, it must be important. But it's so esoteric that I thought people who knew about this sort of thing must be very smart. I wanted to be one of them.

By the time I was in high school, I was already oriented toward research in theoretical physics, and I've never changed that aim. To me, there's something immensely satisfying about the fact that we can work at our desks with abstractions very far removed from ordinary human intuition and still, at the end of the day, come up with scientific theories that say something about the real world: why chemicals behave the way they do, or why other things behave the way they do. That's satisfying in the sense both of giving you a sense of accomplishment and also, in a more childish way, of giving you a feeling of being in on a secret.

Edward O. Wilson

Sociobiology. Pellegrino University Professor, Emeritus, Harvard University.

Most children have a bug period, and I never grew out of mine. One of the things that put me on the path to becoming a specialist in insects, particularly ants, was an accident. At the age of seven, fishing from the shore of Perdido Bay, in Florida, I carelessly yanked my rod too hard with a pinfish on the line. The fish flew out of the water and struck my right eye. One of the pinfish's spines pierced the pupil, and within a few months I lost the sight in that eye.

I already knew that my consuming interest was in animal life. I had to have one kind of animal to study if not another. The attention of my surviving eye turned to the ground. I would thereafter celebrate the little things of the world, the animals that can be picked up between thumb and forefinger and brought close to my good eye for inspection.

Then, when I was 10, my nomadic family moved to Washington, D.C. We lived within easy reach of the National Zoo and the National Museum of Natural History. Here I was, a little kid tuned to any new experience so long as it had to do with natural history, with a world-class zoo on one side and a world-class museum on the other, both free of charge and open every day. I spent hours at a time roaming through both of them. The course of my life was set.

And though that course has led me into work in such broad fields as biodiversity and sociobiology, I placed ants at the center of my professional life. The most important topic I addressed was their means of communication.

It had seemed to me that the fixed-action patterns of ants, living in darkness and underground, must be triggered by chemicals rather than by sight and hearing. My first subject was the fire ant, which is one of the easiest social insects to cultivate in a laboratory. I devised a new kind of artificial nest consisting of Plexiglas chambers and galleries resting on broad glass platforms. The arrangement kept the entire colony in continuous view, allowing me to run experiments and record the responses of the ants at any time.

The most conspicuous form of communication in fire ants is the laying of odor trails to food. I soon found that the ants doing that touched the tip of the abdomen to the ground and dragged it for short intervals along the surface. Other ants followed that trail to the food.

Now I had to locate the source of the chemical. Dissecting abdomens and removing the small organs they contain, I made artificial trails from the contents of those organs. In the end, I found that Dufour's gland carries the chemical that activates the response to food trails. Eventually we identified the chemical as a farnesene.

Since then, many more pheromones, as these triggering chemicals are called, have been discovered. They have opened a new sensory world to biologists. We have come to appreciate the simple fact that most kinds of organisms communicate by taste and smell, not by sight and sound.

Chen Ning Yang

Particle physics. Nobel Prize in Physics, 1957, for the penetrating investigation of the so-called parity laws, which has led to important discoveries regarding the elementary particles. Albert Einstein Professor of Physics, Emeritus, State University of New York, Stony Brook.

Shakespeare likened life to a play of seven acts, of which the seventh and final act

> Is second childishness and mere oblivion;
> Sans teeth, sans eyes, sans taste, sans everything.

I am, fortunately, with teeth, with eyes, with taste, with almost everything, and I have been given the chance to help Tsinghua University to build a research center. Tsinghua University campus is where I had grown up. My life has gone full circle:

> In my beginning is my end,
> . . . .
> In my end is my beginning.

Norton Zinder

Bacterial genes. John D. Rockefeller, Jr., Professor, Rockefeller University.

The days of my childhood began before the existence of antibiotics. Born in 1928, I was a very sickly child; infected ears and sore throats recurred. I had scarlet fever and pneumonia, two serious diseases. My uncle was a doctor, and he treated me. Early on, he gave me neoprontosil, which was the precursor of sulfanilamide. As a curious child, I was interested in the process of these diseases. When I was nine, I used to go with my uncle on his rounds and carry his bag for him. When I was in the Bronx High School of Science, I joined the biology club and learned some laboratory techniques. It was then possible for me to do blood counts for my uncle. I entered Columbia College during the last phase of World War II as a premedical student. I was an assistant to W. Maas, a graduate student, helping him to fractionate eye pigments of drosophila, a genus of fruit flies. Because of the war, college was three semesters a year, and I graduated at the age of 18. Applying only to a few medical schools, I was drowned out by the returning veterans. My adviser could not believe I hadn't gotten into any medical school, and we decided to change my thinking. I applied as a graduate student to the eminent geneticist Josh Lederberg, who was just setting up a lab at the University of Wisconsin.

I arrived in July of 1948 and spent the month getting settled and learning microbiological techniques. On August 8, I did an experiment under Josh's tutelage that opened up bacterial genetics and its biochemistry. Until that time, mutant bacteria that had nutritional requirements could be obtained only by brute force. However, when bacteria were in a minimal culturing medium and had a growth factor requirement, they couldn't grow. This would save them from killing by penicillin, because killing requires growth. Suddenly we had a way to obtain an infinite supply of mutants. They could be used as genetic markers and screens for biochemical pathways used by bacteria. The many mutants helped set me up to do experiments that resulted in the discovery of transduction, which is the transfer of genes from one organism to another by an agent such as a bacteriophage—a virus that infects bacteria. After finishing my Ph.D. at Wisconsin, I came to the Rockefeller University. I have been there since and became a professor in 1964.

Bacteria were once thought to reproduce asexually. It is now known that they have three mechanisms for exchanging genetic information. One is transduction. Another is transformation, which is the addition of naked DNA. The third is conjugation, by which two cells contact each other and exchange DNA. The first two limit the amount transferred, about 1 percent of the genome. Bacterial genes also mutate, providing new genes for evolution. Certain bacteria cause us disease. To treat disease we have a few dozen antibiotic substances which are, in turn, microbial products. The bacteria can, in turn, mutate to resistance to a particular antibiotic or even in some instances half a dozen of them. Since the mutation rate is proportional to the number of organisms exposed, the overuse of antibiotics works to increase the appearance of resistance. There is a real public debate about the use of small doses of antibiotic in animal feed to make a healthier and larger food animal. Many people see this practice as a source of antibiotic-resistant bacteria. The matter has not been settled, but it is clear that the bacteria are in an evolutionary war with the various antibiotics. Resistance that can be transferred sexually certainly provides a potential for the spread of resistance.

Freeman Dyson

Mathematical physics. Professor of Physics, Emeritus, Institute for Advanced Study, Princeton, New Jersey.

My strong suit was always mathematics. I was not driven to become a scientist by any craving to understand the mysteries of nature. I never sat and thought deep thoughts. I never had any ambition to discover new elements or cure diseases. I just enjoyed calculating and fell in love with numbers. Science was exciting because it was full of things I could calculate.

I remember vividly one episode from early childhood. I do not know how old I was. I know only that I was young enough to be put down for an afternoon nap in my crib. The crib had solid mahogany sidepieces so that I couldn't climb out. I didn't feel like sleeping, so I spent the time calculating. I added one plus a half plus a quarter plus an eighth plus a sixteenth and so on, and I discovered that if you go on adding like this forever you end up with two. Then I tried adding one plus a third plus a ninth and so on, and discovered that if you go on adding like this forever you end up with one and a half. Then I tried one plus a quarter and so on, and ended up with one and a third. I had discovered infinite series. I don't think I talked about this to anyone at the time. It was just a game that I enjoyed playing.

In 1931, when I was seven, the asteroid Eros came unusually close to Earth. This is the same asteroid that the spacecraft NEAR (Near Earth Asteroid Rendezvous) explored and landed on in 2001. It is the biggest of the asteroids that regularly come close to us. In 1931, there was much talk of the terrible damage that Eros could do if it should collide with Earth. I listened to my parents talking about this at the breakfast table. My father told me that the Astronomer Royal Sir Frank Dyson was leading an international effort to observe and calculate its orbit precisely. This was important because it would give us a more accurate measure of the distance between Earth and the Sun. Sir Frank was not related to us, but he came from the same part of Yorkshire as my father, and my father knew him. I liked the idea of calculating orbits precisely, and I thought maybe I will one day be Astronomer Royal and calculate orbits too. When my mother died 40 years later, I found among her papers a fragmentary story that I wrote at the age of nine, entitled "Sir Phillip Roberts's Erolunar Collision," about an astronomer who calculates the orbit of Eros and discovers that it is heading for a collision with the Moon. He predicts that the collision will happen in 10 years' time, enough time for him to organize an expedition to the Moon to observe the collision from close by. At that point the story stops. Reading it again today, I noticed that Sir Phillip makes his great discovery by calculating, not by observing.

At age 14, I read the book *Men of Mathematics*, by Eric Temple Bell, a collection of romanticized biographies of great mathematicians. Bell was a professor of mathematics at the California Institute of Technology and also a gifted writer. He wrote with authority about mathematics and knew how to pull at the heartstrings of susceptible teenagers. His book seduced a whole generation of young people to become mathematicians. Although many of the details in the book are historically inaccurate, the important things are true. He portrays mathematicians as real people with real faults and weaknesses. He portrays mathematics as a magic kingdom that people of many different kinds can share. The message for the young reader is "If they could do it, why not you?"

At age 17, I met my first real mathematician, a junior lecturer at Southampton University called Daniel Pedoe. He came to my high school once a week and gave me private lessons. The sessions with Pedoe were a revelation. He had been a research student in Rome, and a member of the Institute for Advanced Study in Princeton, before World War II. He knew personally the legendary figures of mathematics, and he knew the latest fashionable stuff that they had been doing. His own field was geometry. Pedoe and I became lifelong friends. I did not become a geometer, but I acquired from Pedoe a taste for the geometric style that makes mathematics an art rather than a science. Pedoe later became a professor at the University of Minnesota and a leading spokesman for geometry in the international community. He published a book, *Japanese Temple Geometry Problems*, with his friend Hidetosi Fukagawa, about an elegant indigenous version of geometry that flourished in Japan during the centuries of isolation from Western influences.

After World War II came to an end, I needed to leave England and see the world. After six years of war, everyone wanted to travel. I applied for a Harkness Fellowship to spend a year in America and was one of the lucky winners. At the Cavendish Laboratory in Cambridge, I happened to meet Sir Geoffrey Taylor, an expert in fluid dynamics who had been at Los Alamos during the war. I asked Taylor where I should go in America. He said without any hesitation, "Oh, you should go to Cornell, that is where all the brightest people from Los Alamos went after the war." He said Hans Bethe, who had been head of the theoretical division at Los Alamos, was the

person I should work with. He knew Bethe well and would put in a good word for me. I knew almost nothing about Cornell, but I took Taylor's advice. I went to Cornell to work with Bethe. Bethe was just what I needed. He gave me a difficult but not impossible problem to solve. After I solved it, I really belonged to the club.

My final stroke of luck was meeting Richard Feynman. Feynman was a young professor at Cornell, not yet famous. I had never heard of him before I came to America. He was rebuilding the whole of physics from the bottom up, using a geometrical language with diagrams that nobody except Feynman understood. I recognized that Feynman was a genius and my job was to understand his language and explain it to the world. So that is what I did. I spent as much time as I could with Feynman. I watched him drawing diagrams on the blackboard and listened to him talking. He liked to go for long walks and talk about everything under the sun. After a year at Cornell, I understood Feynman's way of thinking and translated it into the old-fashioned mathematics that I had learned in England. I published two papers explaining why Feynman's methods worked. My papers were best sellers, and Feynman's language became the standard language of particle physicists all over the world. At the age of 25, I was famous. At a meeting of the American Physical Society where I was one of the main speakers, Feynman said to me, "Well, Doc, you're in." Childhood was over, and I was free to spend the rest of my life finding problems in various areas of science where a tablespoonful of elegant mathematics could make a big difference.

I belong to the majority of scientists who practice science as a useful skill like house building or cookery, not to the minority who practice science as a philosophical inquiry. I have never cared whether the problems I was trying to solve were important or unimportant. The unimportant problems in pure mathematics are just as much fun as the important problems in atomic physics or biology. I am one of the luckiest people on Earth, being paid for doing what I enjoy most. I don't pretend to understand why I am so lucky. It just happens that I was born fluent in the language of mathematics and that mathematics is nature's language too.

Acknowledgments

This book would not exist without the inspiring enthusiasm, wisdom, and knowledge of Gerard Piel. I am also grateful to Howard Stein, whose passion for both photography and science played a formative role in developing my initial interest in photographing scientists. Torsten Wiesel kindly suggested additional subjects, whom I am thankful to have been able to include. James T. Rogers was extremely helpful in working on the texts with me. Safi Bahcall also made possible several portraits which add significantly to the book. My agent, Timothy Seldes, found me just the right publisher and editor, Norton and Jim Mairs, who have been nothing less than marvelous at every juncture. It has been a pleasure to work with the designer, Katy Homans. My assistant, Trellan Smith, continues to be my "right arm."

Alphabetical Index of Scientists

(Dates and places refer to when and where the scientists were photographed.)

Eric Lander, page 94
August 24, 2001, Santa Cruz, California

Nicole Le Douarin, page 96
November 12, 2003, Paris, France

Richard Leakey, page 98
April 3, 2003, New York City

Leon Lederman, page 92
October 13, 2003, New York City

Andrei Linde, page 102
March 21, 2003, Stanford, California

Susan Lindquist, page 104
October 10, 2003, Cambridge, Massachusetts

Roderick MacKinnon, page 106
March 29, 2004, New York City

Geoffrey Marcy, page 110
March 24, 2003, San Francisco, California

Lynn Margulis, page 112
March 31, 2003, New York City

Ernst Mayr, page 114
May 6, 2003, Cambridge, Massachusetts

Maclyn McCarty, page 68
January 23, 2004, New York City

Matthew Meselson, page 120
May 7, 2003, Cambridge, Massachusetts

Philip Morrison, page 118
February 4, 2003, Cambridge, Massachusetts

Christiane Nüsslein-Volhard, page 124
March 12, 2004, Frankfurt, Germany

Leslie Orgel, page 122
March 12, 2003, La Jolla, California

Wolfgang Panofsky, page 126
March 21, 2003, Los Altos, California

Ruth Patrick, page 82
October 7, 2003, Philadelphia, Pennsylvania

C. R. Rao, page 128
October 2, 2003, University Park, Pennsylvania

Martin Rees, page 132
May 19, 2003, Cambridge, England

Colin Renfrew, page 134
November 13, 2003, Cambridge, England

Miriam Rothschild, page 136
May 12, 2001, Peterborough, England

Vera Rubin, page 116
October 16, 2003, Washington, D.C.

Allan Sandage, page 138
March 27, 2003, Pasadena, California

Frederick Sanger, page 140
December 9, 2003, Swaffam Bulbeck, England

Maarten Schmidt, page 142
March 13, 2003, Pasadena, California

William Schopf and
Jane Shen-Miller Schopf, page 58
March 14, 2003, Los Angeles, California

John Schwarz, page 144
March 27, 2003, Pasadena, California

Steven Stanley, page 146
October 7, 2003, Baltimore, Maryland

Joan Steitz, page 148
October 9, 2003, New Haven, Connecticut

Sir John Sulston, page 152
March 21, 2004, Cambridge, England

Lynn Sykes, page 150
April 1, 2003, New York City

Ian Tattersall, page 154
April 14, 2003, New York City

Harold Varmus, page 156
October 8, 2003, New York City

James Watson, page 160
December 19, 2003, New York City

Robert Weinberg, page 158
October 10, 2003, Cambridge, Massachusetts

Steven Weinberg, page 162
February 2, 2004, Austin, Texas

Torsten Wiesel, page 100
September 24, 2003, New York City

Edward O. Wilson, page 164
November 23, 2004, Cambridge, Massachusetts

Chen Ning Yang, page 166
March 3, 2003, St. James, New York

Norton Zinder, page 168
October 8, 2003, New York City

Marek Zvelebil, page 108
May 22, 2003, Sheffield, England